Currere
AND THE
Environmental Autobiography

A Book Series of Curriculum Studies

William F. Pinar
General Editor

Vol. 4

PETER LANG
New York • Washington, D.C./Baltimore • Bern
Frankfurt am Main • Berlin • Brussels • Vienna • Oxford

Marilyn N. Doerr

Currere AND THE Environmental Autobiography

A Phenomenological Approach to the Teaching of Ecology

PETER LANG
New York • Washington, D.C./Baltimore • Bern
Frankfurt am Main • Berlin • Brussels • Vienna • Oxford

Library of Congress Cataloging-in-Publication Data
Doerr, Marilyn N.
Currere and the environmental autobiography: a phenomenological approach
to the teaching of ecology / Marilyn N. Doerr.
p. cm. — (Complicated conversation; vol. 4)
Includes bibliographical references.
1. Ecology—study and teaching. 2. Education—Biographical methods.
3. Phenomenology. I. Title. II. Series.
QH541.2.D64   577'.071—dc21   2003040090
ISBN 0-8204-6369-8
ISSN 1534-2816

Bibliographic information published by **Die Deutsche Bibliothek**.
Die Deutsche Bibliothek lists this publication in the "Deutsche
Nationalbibliografie"; detailed bibliographic data is available
on the Internet at http://dnb.ddb.de/.

Cover photo by Hal Eastman
Cover design by Lisa Barfield

The paper in this book meets the guidelines for permanence and durability
of the Committee on Production Guidelines for Book Longevity
of the Council of Library Resources.

© 2004 Peter Lang Publishing, Inc., New York
275 Seventh Avenue, 28th Floor, New York, NY 10001
www.peterlangusa.com

All rights reserved.
Reprint or reproduction, even partially, in all forms such as microfilm,
xerography, microfiche, microcard, and offset strictly prohibited.

Printed in the United States of America

To Dick, Chris, Mark, Tony

How important it is to walk along, not in haste but slowly, looking at everything and calling out

Yes! No! The

swan, for all his pomp, his robes of glass and petals, wants only to be allowed to live on the nameless pond. The catbrier is without fault. The water thrushes, down among the sloppy rocks, are going crazy with happiness. Imagination is better than a sharp instrument. To pay attention, this is our endless and proper work.

—Mary Oliver, from "Yes!No!"

TABLE OF CONTENTS

Chapter I
Introduction: A Fan of Eyes ... 1

PART ONE: A Science Curriculum Enhanced by Pinar's *Currere* ... 5

Chapter II
Currere ... 7

Chapter III
Currere in My Classroom ... 20

Chapter IV
Data from the First Year's EAs: Discovering "Nature's Temporary Cushion" ... 32

Chapter V
Narratives from Subsequent Years' EAs: "Not Till We Are Completely Lost or Turned Around Do We Appreciate the Vastness and Strangeness of Nature" ... 72

Chapter VI
My Pedagogy ... 95

Chapter VII
What Kind of Ecologists/Scientists Are Schools Turning Out? ... 114

PART TWO: A Science Curriculum That Is Both Phenomenological and Postmodern ... 129

Chapter VIII
An Ecology Curriculum That Is Both Phenomenological and Postmodern ... 131

Chapter IX
Caring in Schools ... 140

Chapter X
Insecurities/Gender Issues 147

Chapter XI
Moving Away from Egocentrism 158

Chapter XII
Politicization 168

Chapter XIII
Definitions of Success 177

Chapter XIV
Conclusion 182

Appendix
How I Did My Research 187

Bibliography 197

Chapter I
Introduction: A Fan of Eyes

> Why is it so difficult simply to capture what was there? That old problem of trying to show things both sequentially and simultaneously. If I drew that scene I'd show everything happening all at once, everyone present and every place visible, from the bottom of the river to the clouds. But when I describe it in words one thing follows another and everything's shaped by my single pair of eyes, my single voice. I wish I could show it as if through a fan of eyes. Widening out from my single perspective to several viewpoints, then many, so the whole picture might appear and not just my version of it. As if I weren't there.
> —Andrea Barrett, *The Voyage of the Narwhal*, 26–27

From my office window tonight I can see Mars. As it makes its closest path to us in the last 50,000 years, I sit here and think of Mars as Other. Is it our Other place? For so long we have romanticized Mars as this reddened orb that maybe once housed life and water and, possibly, culture. Now maybe we need to think more seriously about using this space because what is happening on our own planet is getting scary. The glaciers in the Andes are shrinking fast—too fast; no one is prepared to try to rectify the situation. Soils are badly salinized from overuse of chemical fertilizers. Childhood illnesses are becoming increasingly connected to the environment. There are still too many people with unsafe water to drink and unhealthy air to breathe. E. O. Wilson would call this our "environmental bottleneck" (Wilson 1998, 288). When we depend on artificial means to keep the planet alive, we "render everything fragile," and "impoverish our own species for all time" (298). How do we move through this bottleneck we have produced for ourselves? This book is about a very small attempt to find hope in the seemingly hopeless desecrations of the earth, and, maybe, push through.

I work at a private all-boys college preparatory school. The school houses approximately one thousand boys, grades K-12, on two distinct campuses. The campus where I work, grades 9-12, consists of 234 acres, much of it wooded and pristine. There is a large lake at the back of the school building. During the 1998–99 school year I taught an ecology class on this campus, the first time an ecology class had been taught at the school. I had two sections, with ten students in each section. The classes were kept deliberately small so I could use available outdoor equipment, such as boats and nets, and allow everyone to participate; I did not want to have a class where I did all the demonstrating and the students merely watched. The twenty students were almost all seniors taking this course as a science elective. Most had been through the traditional sequence of freshman year biology, sophomore year chemistry, and junior year physics. A few students were taking the course as part of their requirement of three years of science, having opted out of the physics course their junior year.

I have now been teaching that ecology class for five years. I have taught the class to more than 135 students, most of them seniors. Using an adaptation of William Pinar's pedagogy of *currere,* I attempt to teach an ecology course that I feel will not only appeal to eighteen-year-olds but that will instill in them a philosophy of living with ecological principles.

In a world such as we are in now, less and less value is being placed on creative teaching and more emphasis is being put on prescriptive teaching, teaching where schools can publicly laud their students for passing various proficiency tests in reading, writing, math, and so on. When a pedagogy such as the one about to be explained herein is addressed to the majority of practicing teachers, it is not really skepticism that prevents most from even trying something like it. It is the atmosphere created in today's schools as accountability becomes the watchword of the day, and taxpayers and parents demand certain levels of writing and reading and math deemed appropriate for certain grade levels. The teacher who advocates methods such as the one about to be presented here may be super-enthusiastic about the outcomes of such an experiment, but s/he will find spirits quickly dampened by the all-pervading milieu of national norms, national testing, and national curriculum. Where does the art of pedagogy fit into this early part of the twenty-first century? In many ways, it does not.

In 1993, *Benchmarks for Scientific Literacy* was published by the Department of Education. Annette Gough comments that the focus of these benchmarks was on "what scientific literacy is *about* but contained little discussion about what that literacy might be *for"* [emphasis in original] (Gough 1998, 202). She goes on,

> There is a large body of research which indicates that simply acquiring seemingly relevant scientific knowledge provides no guarantee of later involvement in scientific decision making, and this then raises questions about the actual likelihood of achieving scientific literacy through a focus on scientific content in curriculum. (202)

I have seen this over and over again in my years of teaching. The students with the highest scores in the AP science classes often care little about science issues in politics or the ethics of advancements in science as regards society. It is not enough to know the light and dark reactions in photosynthesis or all the steps of the Krebs cycle. What is important is how that material relates to the way scientists and others use that knowledge for the betterment of their own and others' lives. I believe the *currere* practice detailed in this book personalizes scientific concepts and helps develop understanding of the stories of which we play a part.

If I had my way, every student in the country would take a course in ecology when s/he turns eighteen.[1] It is then that the student can begin to internalize ecological principles into his life and make some conscious decisions about how he wants to live that life. We educators teach about recycling, rainforest production and destruction, basic water quality, and pollution in earlier grades, but it isn't until the student is in late high school that he begins to realize that these issues impinge on his life. The student's earlier groundings in ecology are important and give good background to the study of ecology, but for most students, this is all they will ever have in schools in any kind of formal curriculum in ecology. By college, very few students will continue their education in science, let alone in ecology, one of the newest sciences.

I am convinced that the teaching of ecology is not internalized sufficiently unless there is an emotional component to the curriculum. The student must make connection with the space in which he lives in order to make philosophical changes in his life to protect that space. When I first learned that I was to teach two sections of an ecology elective to senior boys, I was determined to find some way to move my students from "I know" to "I care" to "I want to do something about this," what I would consider a critical pedagogy of ecology. When I came across William Pinar's *currere,* I knew I had found my first "experiment" for the course.

---

[1] From this point on in the paper I will be using male pronouns because I am working in an all boys school. That is the only reason. I certainly feel that everything stated herein is equally applicable to females.

PART ONE

A SCIENCE CURRICULUM ENHANCED BY PINAR'S *CURRERE*

Chapter II
Currere

> The Great Work before us, the task of moving modern industrial civilization from its present devastating influence on the Earth to a more benign mode of presence, is not a role that we have chosen. It is a role given to us, beyond any consultation with ourselves. We did not choose.
> —Thomas Berry, *The Great Work*, 1999

Philosophy of *Currere*

What exactly is *currere*? In order to understand how *currere* was able to help my students become more involved ecologists, it is necessary to have some basic knowledge about *currere*. This chapter deals with *currere* and its history.

When I studied curriculum as an undergraduate, I was taught that the purpose of curriculum was to transmit the culture. The culture was to be passed on, mostly unchallenged, so that status quo could be maintained (Kliebard 1995; Pinar, Reynolds, Slattery, and Taubman 1995, 69–158). There was no thought of pedagogical sites other than the classroom, no consideration of the political and cultural aspects of the hidden curriculum. It was all too simplistic. Transmitting the culture didn't consider the role of the student. Where was the idea of agency? Culture was not considered a social construction that could be changed. The idea that culture could be challenged and critically discussed was seldom addressed (Giroux 1997, 3–34; Pinar et al. 1995, 159–240). I knew I was no automaton, and certainly there were many, many students just like me. If a curricular theorist pretended that defining a course of study was sufficient for curriculum development, how could schools ever become sites of critical pedagogy? Curriculum theory had to be more than simply developing a course of study. When I learned of William Pinar and his conception of curriculum theory, I knew I had found a soulmate.

Pinar's ideas of curriculum theory are encapsulated in *currere*. *Currere* focuses on the educational experience of the individual as reported by the individual; it seeks to describe what the individual himself/herself makes of behaviors. The concept of *currere* is grounded in existentialism, phenomenology, Jungian psychoanalysis, the psychiatry of R. D. Laing, and aspects of literary and educational theory (Pinar 1975a, 403).[1] Pinar characterizes the

---

[1] Pinar taught English in high school before going on to become a professor of education and curriculum theory, and frequently uses literary analysis as examples for his theory.

method[2] with the following four steps: the method is (1) regressive, involving description and analysis of one's intellectual biography; (2) progressive, involving a description of one's imagined future; (3) analytic, calling for a psychoanalysis of one's phenomenologically described past, present, and future; and (4) synthetic, totalizing the fragments of experience and placing this integrated understanding of individual experience into a larger political and cultural web (Pinar, 1975a, 404). (See chapter 3 for my variation on this method.)

Pinar's use of *currere* seems analogous to Johannes Itten's use of the Basic Course with the Bauhaus, where Itten developed ways for his students to explore and become more aware of their experience through physical exercises and through meditation. This helped them experience their own bodies in the present. In allowing students to locate their own current intellectual interests in their life histories, both Itten and Pinar joined content and form into a powerful, postmodern curriculum (Pinar et al. 1995, 578).

Sartre's *Search for a Method,* in which he explores the connection between existentialism and Marxism, intrigued Pinar to press for a way curricularly to find Husserl's transcendental ego (Mohanty 1997, 36, 44), a place where one experiences a continuity in educational experience, where normally most of us experience discontinuity and disjointedness. Pinar asked questions such as, What is the relation between the individual and the collective, between self and other, between internal and external? How does context influence self and vice versa? (Pinar 1975a, 407). Through a directed free association, he began to think about his own early experiences with school. He began to gather data that he felt were phenomenologically accurate as he relived early and present experiences uncritically. Only when he gathered sufficient data did he begin to analyze. He asked himself questions such as, What correspondences do I find between conditions in my private life and those in my academic life? What seem to be characteristic responses to various teachers, to various forms of work? (409). This process, he stated, will lead to "a generalized inner-centeredness and hopefully initiate or further the process of individuation, leading to the gradual formation of the transcendental ego" (410).

Pinar wanted to change the emphasis of curriculum theory from one of prescription and guidance to one of understanding. This involved a shift in perspective, both cognitively and affectively. It would be a process of turning inward, of changing consciousness. Praxis, purposeful human activity, would

---

[2] The word "method" has taken on a negative connotation with today's educators because of its link with prescriptive teaching. Pinar himself calls *currere* a method (for example, the title of chapter 4 in *Toward a Poor Curriculum* is "The Method"), and I continue to use the word in this book to show the procedural nature of *currere*. *Currere* is in no way prescriptive and deskilling for teachers, but there are stages to the process that I feel are important for its use.

be affected. He called this change *currere,* from the Latin infinitive of "to run," to show the active nature of curriculum. He viewed *currere* as the kernel, or core, of a reconceived and revitalized curriculum theory field (Pinar 1975a, 413).

In *currere,* the individual is stressed. The focus is not on a defined curriculum per se but on the individual's experiences encountering that curriculum. The method is self-hermeneutical and phenomenological (Pinar 1975a, 403).

In Camus's *The Plague,* the world of the protagonists is infected with a plague, a sickness that Camus uses metaphorically as the sickness of indifference. There is a weariness to life that obscures any connection an individual may be making with his space. But, according to Camus, one can "refuse" to be attacked by the plague. Tarrou, one of the main characters, says this refusal must be met with constant vigilance:

> [W]e must keep endless watch on ourselves lest in a careless moment we breathe in somebody's face and fasten the infection on him. What's natural is the microbe. All the rest—health, integrity, purity (if you like)—is a product of the human will, of a vigilance that must never falter. The good man, the man who infects hardly anyone, is the man who has the fewest lapses of attention. (Camus 1948, 229)

I believe it is this vigilance that Pinar is addressing with *currere*. It is an awakening to the indifference that is so often promulgated in schools, where students absorb the reality the schools construct. Their consciousness becomes submerged in the oppressive atmosphere of the classroom. Pinar believes the habits and compliance that schools produce can be shaken up by allowing students to address questions that increase awareness of how they live within their worlds. Tarrou becomes self-reflective as Camus's story continues. He tells the story of his own life, and it is in the recalling of past experiences that he begins to understand the meaning of the plague. He uses this retelling of his past to figure out how he has constituted his world, formed his own traditions, and provided some context for his future. When he is asked what path he would follow to attain peace, he replies, "The path of sympathy." He could only make such a statement after making a deliberate attempt at articulating his mental life in the context of human-being-within-the-world. It is this type of immersion that Pinar is calling for in *currere*.

Teachers, Grumet says, must refuse the mechanical life of schools. They must generate knowledge structures linked to relevant themes. They must bring consciousness back to the foreground of educational experience (Pinar and Grumet 1976, 111–47).

Paulo Freire talks about consciousness when he discusses his idea of praxis. To Freire, praxis is a method of problem posing and problem solving where the outcome transcends an existing social situation. "Liberation is a praxis, the action and reflection of men upon their world in order to transform it" (Freire 1970, 60). This liberation cannot take place until the student is aware of where he stands within this world he wishes to change. It demands a self-reflection whereby the student analyzes his relationships within that space. Why does he feel like he does? Why does he think like he does? Why is he drawn more toward one argument than another? Answering questions such as these is part of *currere*.

Dwayne Huebner writes about the language of curriculum and how the vigilance of which Camus was writing is needed by educators if one is not to get swept into the reproductive nature of most schooling today. There is a danger of being socialized into existing institutions (Huebner 1974, 36). If education is to be intentional, he says, we must be aware of three things, which he calls "three facets of man's temporality": (1) the phenomena of memory and traditions as these store and make accessible the past; (2) the activity of interpretation—the art of hermeneutics—the linkage among past, present, and future; (3) the phenomenon of community as a caring collectivity in which individuals share memories and intentions (37). If a school fails to recognize and encourage these facets, it does a disservice to its students. Schools are in constant conflict over the individual needs of students and the needs of society. Yet schools hide this conflict and give the illusion that it does not exist. This conflict, I believe, is part of what Pinar calls the need for Reconceptualization (Pinar 1975c). Using Huebner's three facets—memory, interpretation, and the link between individuals and society—he calls for a different language in which this conflict would become part of the curriculum. Education should not be about reproduction, but should instead be part of Freire's praxis where society becomes transformed and liberated.

"The path of sympathy," which Camus refers to as a pathway to peace, is very much needed in dealing with the global problems that are threatening our survival. Overpopulation, exhaustion of energy resources, pollution of the biosphere, the threat of nuclear weapons, dehumanizing structures in technology, and the widening gap between rich and poor share a complex causality, but I believe they share a common cause: an indifference to human values. Honesty, integrity, cooperation, responsibility, justice, caring, self-fulfillment, joy—these are values that become moral imperatives in a global world.

Through the process of *currere* the student becomes aware of his perception of his connectedness to others. He begins to explore how human values fit into his life and the degree of importance he has so far delegated to them.

History of *Currere*

At the 1967 Curriculum Theory Conference at Ohio State University, Dwayne Huebner introduced phenomenology to curriculum studies for the first time (Pinar et al. 1995, 417). Phenomenology is interested in the process of living and in an individual's or group's reactions and attitudes to behavior. Huebner suggested that the lived world of both teachers and students gets lost in traditional curriculum. He felt schools were doing a disservice to their students by not valuing lived experience (Huebner, in Pinar 1975, 237–49).

By 1972, Pinar had begun to understand the new movement in the field of curriculum theory spurred by Huebner's call for phenomenological experience as an essential component of an education. He felt the time was ripe for what he and others were calling the Reconceptualization, a new paradigm for curriculum theory. With the help and support of his mentor, Paul Klohr, Pinar would organize a curriculum conference for 1973 that featured papers and presentations by many of the theorists taking part in this reconceptualization of curriculum theorizing (Pinar 1974b). No longer was curriculum to be simply a course of study designed to transmit the culture.

In 1976, he and Madeline Grumet published *Toward a Poor Curriculum*, the seminal work in which he would introduce *currere*. Patrick Slattery, in his *Curriculum Development in the Postmodern Era* (1995), notes that both Pinar and Grumet were influenced by the stream-of-consciousness writings of James Joyce, Marcel Proust, Virginia Woolf, and William Faulkner and the paintings of artists such as Jackson Pollock (Slattery 1995, 55).[3] That influence is apparent not only in the method of *currere* but in the style in which *Toward a Poor Curriculum* was written. Unfortunately, curricular theorists and practitioners were not receptive toward the book in 1976 (Pinar 1994, 1), and it took almost twenty years for the curricular world to recognize the genius in this work (Pinar et al. 1995, 515–16).

By the end of the 1970s Pinar had characterized three schools of curriculum theory: the traditionalists, the conceptual-empiricists, and the reconceptualists (Giroux 1981; Kliebard 1995; Pinar 1978,). The traditionalists saw curriculum theory as "service to practitioners" (Pinar in Giroux 1981, 88). The goal was to design a system that achieved an educational end and not to attempt to explain an existing phenomenon (Tyler in Schaffarzick and Hampson 1975, 18). Their work focused on helping the teachers, and there-

---

[3] Jackson Pollock appears to be a particular favorite of Pinar's. It is interesting that Carl Djerassi, in writing the science-based novel *The Bourbaki Gambit*, uses Pollock's work to produce one of the best descriptions of complexity theory I have read. "A Jackson Pollock," Djerassi writes, "looks very messy—all those drips and blobs. But its complexity is actually the kind we call simple: there's a lot going on there, but it's made up of essentially simple forms" (Djerassi 1994, 146).

fore the school, perform optimally and efficiently. Tools were designed—plans of study, developmental progressions, assessments, innovative schedules—to help educators get ready for Monday morning. Traditional curriculum theory was ahistorical and patterned after the model of the natural sciences (Giroux 1981, 99). Pinar hypothesized that approximately 90 percent of the people working in the field of curriculum studies were traditionalists. (He no longer feels that way today—personal communication with him, Dec. 1999.) Ralph Tyler, with his theories of scientific management, was their guru, and rationality was their password. Going all the way back to the ideas of Horace Mann of schooling designed to produce workers for capitalism in specific economic and class structures, traditionalists continued to see the school as a factory, producing workers for the existing economy. Traditional curriculum theorists' work was largely prescriptive.

Conceptual-empiricists saw education as part of social science research. They took their name from the social science and anthropological research methods they employed, developing hypotheses to be tested (Pinar in Giroux 1981, 91). Conceptual-empiricists recognized that most traditional curriculum theory was never fully played out in a classroom. Teachers took what they liked about a curricular change and modified it to suit their particular styles and objectives. Conceptual-empiricists studied how educators viewed change in a classroom and how administrators used the dialectic of stability and change to run their schools. Conceptual-empiricists may have been studying ahistorical problems, but they historicized them as part of their methodology and attempted to make generalizations, arguing inductively. While stating that their work was apolitical, they generally maintained the status quo.

Reconceptualists saw research as more than an intellectual act; it was unavoidably political. Reconceptualists looked at the culture at large and argued that educational problems could not be solved until the larger cultural issues were solved. There was a commitment to a comprehensive critique and an emphasis on emancipation. Reconceptualists did not see curricular changes as a simple "plugging in" to the already existing traditional curriculum, making small changes here and there. There had to be fundamental change—hence the term "reconceptualization": "to conceive again."[4]

---

[4] I, however, am using a "plugging-in" approach with the EA (Environmental Autobiography). If I was going to experiment with *currere*, a method that comes from the reconceptualist camp, I could not see any other way. I had to work within the extant order. I am working in schools twenty-five years after the reconceptualists held their first conference and still do not see examples of their work in American schools. If *currere* is effective, I would argue that it can be effective within the existing social order. Sometimes "baby steps" are the best we can do.

In 1995, when Pinar, along with Reynolds, Slattery, and Taubman, produced *Understanding Curriculum,* a monumental work that attempted to put the entire history of curriculum theory in focus, Pinar changed this delineation of schools of curricular thought into a broader and less reified structure. "Boundaries are porous" (Pinar et al. 1995, 51), he wrote, and it is seldom that any curricularist is a member of only one of these schools; there is always overlapping. His schools of contemporary curriculum discourse, as outlined in *Understanding Curriculum,* fall under the following nine headings: curriculum as political text; as racial text; as gender text; as phenomenological text; as poststructuralist, deconstructed postmodern text; as autobiographical/biographical text; as aesthetic text; as theological text; and as institutionalized text. (*Currere* is placed under curriculum as autobiographical/biographical text, but it certainly has elements of gender and poststructuralist text.)[5]

Madeleine Grumet, who has worked closely with Pinar, talked about her ideas of curriculum theory in the preface to *Bitter Milk: Women and Teaching*. She said that curriculum "expresses the desire to establish a world for children that is richer, larger, more colorful, and more accessible than the one we have known" (Grumet 1988, xii). She saw curriculum theorists as working with a dialectical model of the way curriculum mediates between person and world. This made curriculum tentative and provisional, a negotiation between the lives we are capable of living and the ones we do live (xiii). Curriculum must call forth individual experience and value it. If we do not use our personal experiences, Grumet said, we risk turning away from the "places where we were most thrilled, most afraid, most ashamed, and most proud…our experience gathers up its convictions and its questions and quietly leaves the room" (xvii). This is where *currere* can be so effective. It pulls experience into the picture and gives it voice.

Pinar extended his critique of schooling as a dehumanizing process of socialization into a method for reading and writing texts of educational experience where the daily experience of students and teachers gave alternative visions for schooling. In his work with Grumet on autobiographical texts of educational experience, he substituted hermeneutics for the positivism of the social sciences that has dominated curriculum research. Grumet wrote, "As we concur with Richard Rorty's claim that hermeneutics has displaced epistemology by providing another politic for scholarship that refuses the dichotomization of subject and object, we invite endless problems of interpre-

---

[5] I do not want to abandon the word "reconceptualization," though, in this book, because I think it points up the idea of new, revolutionary ways of thinking about curriculum. All the fields of contemporary curriculum theorizing ask schools to rethink old paradigms and look at curriculum as more than a chronology and collection of lesson plans.

tation, not as impostors at the banquet planned for the truth but, indeed, as the guests of honor" (Grumet 1988, 60). What a beautiful way of looking at students' work! In 1975 Pinar and Grumet, along with two assistants, worked with eleven student teachers in a teachers' training seminar at the University of Rochester and experimented with *currere* for the first time (Pinar and Grumet 1976, 147–77). They were using *currere* to examine the students' responses to their own educational experiences so the teachers-in-training could see for themselves the baggage they would bring with them into their own classrooms.

Another binary that concerned Grumet was the figure/ground relationship. She utilized *currere* in theatre workshops to help concretize her opposition to this dichotomy. In the late seventies, drawing from the works of theoretician and theatre director Jerzy Grotowski, she asked her theatre students to strip away their "life masks" and allow figure and ground to meld into each other (Pinar et al. 1995, 589). *Currere*, in combination with theatre, allowed her students to become conscious of their own bodies, feelings, thoughts, and words. She sensed a strong link between the two; whereas autobiography was postreflective, reflecting on reflective self-representations, theatre was preflective, preceding the reflective self-representation in actions requiring an immediate, nonverbal response (Pinar and Grumet 1976, 80). Grumet used mime to highlight this link. "The givens of the physical world, the hardness of a floor . . . and the givens of our bodies, their capacities for torsion, stretch, balance, were fused by our intentions into meaningful actions" (quoted in Pinar et al. 1995, 597). The duality of figure/ground, of mind/body, dissolved in this pedagogical mix.

Grumet introduced *currere* into her theatre classes, she said, in order to *protect her students from the curriculum* [my emphasis]. She saw that *currere* was able to keep her students from losing their individual identities in the forms of the plays and their characters. She wanted her students to find their own forms, their own deconstructions of the theatrical experience, and produce their own readings of the world (Pinar et al. 1995, 596). She knew that the stories collected in the students' regression and progression stages would be largely mundane and only occasionally extraordinary, but that as they moved into analysis and synthesis, they would be transformed by the experience (597).

Grumet writes that in the production of autobiography, categorical meanings are suspended wherever possible in the composing process. In the narrative, the autobiographer is seeking not an illustration of categories but the dialectical interplay of experience in the world and ways of thinking about it. In this way, she writes, "the literary narrative that is autobiography resembles the social event that is curriculum: Both function as mediating forms that gather the categorical and the accidental, the anticipated and the unexpected, the individual and the collective. The gap of error or surprise that erupts in the midst of the well-made text is what deconstructionists seek, not to embarrass the author of the erratum but to demonstrate that the power of the person, the text, the meaning, is spurious when we impute to it an utterly consistent, exclusive, bounded, and delineated logic" (Grumet 1988, 67). Grumet writes:

> Using autobiography as a curricular tool implies the curricularist is concerned with making meaning out of knowledge. How is knowledge learned contextualized by the student? What knowledge is worth knowing to this particular student? Pinar feels that curriculum, whether traditional or reconceptualist, expresses value; it reorders experience so as to make it accessible to perception and reflection. Curriculum should relate to the rhythms of history and biography and to the lived spaces of home, work, and community structures. In one of his earliest works on reconceptualist curriculum, Pinar says, "Just as the artists' canvas holds virtual, not actual space, the composer's score, virtual not actual time, the curriculum provides virtual not actual experiences, embodied in the academic disciplines. As such it is a field of symbols, abstract in the science, particular in the arts, for contact with the world. (Pinar and Grumet 1976, 77)

But to bring what we know to where we live has seldom been the project of curriculum, for schooling often has functioned to repudiate the body, the place where it lives, and the people who care for it. There is a distance between what we know and how we live. Philosopher Maurice Merleau-Ponty refers to the "body-subject," a concept he feels is imperative to thinking, the idea that we cannot discount the body when discussing consciousness (Merleau-Ponty 1962). Feminist educator bell hooks refers to the body-mind connection and also sees the link as impossible to separate (hooks 1994, 136–37). Pinar concretizes this concept in *currere*.

It is not only educators who see the advantage of autobiography. Joan Didion, in *Slouching Towards Bethlehem* (1961), a collection of essays on her writing life, writes about how she keeps a notebook with her at all times, where she refers to the distinction between *what happened* and *what it means to me*; she sees her notebook as bringing consciousness to the data she is collecting for her work. The "I" of autobiographical consciousness is an index to a subjectivity that is always open to new possibilities of expression and realization. The "I" is the location of a stream of possibilities.

bell hooks writes about the use of personal experience in liberatory pedagogy: "different, more radical subject matter does not create a liberatory pedagogy . . . a simple practice like including personal experience may be more constructively challenging than simply changing the curriculum" (hooks 1994, 148). When one speaks from the perspective of one's immediate or past experiences, an atmosphere is created in the classroom that allows the students to claim a knowledge base from which they can begin to build. It is a *coming to voice*—a voice that, once established, permits students to speak freely and easily about other subjects. hooks feels also that once students establish that voice they become better listeners; in recognizing their own uniqueness, they become curious about others' (151). There is a respect, then, that is felt throughout the space of the classroom, a respect for each person in the room as, together, they share experience.

The Four Stages of *Currere*

There are four stages in *currere*: regression, progression, analysis, and synthesis.[6] Pinar calls these four steps a "developmental point of view." It is indeed developmental—*currere* is effective only as one moves through each of the four stages in the order Pinar designated. These four steps were designed by Pinar to try to make sense of how formal schooling contributes to our sense of who we are. He writes, "I am taking as hypothesis that I am in a biographic situation, and while in certain ways I have chosen it (and hence must bear responsibility for it), in other ways I can see that it follows in somewhat causal ways from previous situations" (Pinar 1994, 19). He makes a further hypothesis: He does not know how his educational experiences have affected him (20). By recording his educational experiences, he writes about them as he sees them, not as he wants to see them. By placing them into text, visible on a sheet of paper, he can distance himself from the experience and view it from a different perspective. As he accumulates these written experiences, his perspective keeps changing, "because the problem [the experience] is inherently a partial product of my conceptual apparatus, and because this apparatus functions slightly differently since its operator has moved slightly, the problem itself poses itself differently, and hence the problem is different" (21).[7]

Stage One is regression. Regression looks at the past. In regression, the writer records past experience and observes it. There is no analysis here, just a recording of these past observations, a "gathering of data," if you will. Since the focus is educational experience, the observations emphasize life in schools, life with teachers, life with books. The writer is to observe as far back into the past as possible. By recording these observations in a visual text, the writer brings the past to the present. "The words coalesce to form a photograph. Holding the photograph in front of oneself, one studies the de-

---

[6] Pinar delineates the method of *currere* in the fourth chapter of *Toward a Poor Curriculum* (1976) and again in the third chapter of *Autobiography, Politics, and Sexuality* (1994). Because he has changed the text somewhat in the 1994 reprint, I am mainly citing from that chapter.

[7] I found that I was not able to truly understand that sentence until I got through the first year of working with *currere*.

tail, the literal holding of the picture and one's response to it, suggestive of the relation of past to present" (Pinar, 1994, 24).

Stage Two is progression. Progression involves the future. Pinar suggests a meditative posturing here. The writer imagines the future in a free-associative manner. On several occasions, the writer is to put himself in a meditative state where he can allow his mind to see himself a year from now, ten years from now, thirty years from now. He records these observations. The writer looks "at what is not yet the case, what is not yet present.... the future is present in the same sense that the past is present. It influences, in complicated ways, the present; it forms the present" (Pinar 1994, 24).

Stage Three is analysis. Analysis looks at the present. What today is the writer's institutional life as regards schools? What is his intellectual life? What areas of study attract and repel? Another photograph is created. At this point in Analysis, the writer has produced three photographs of himself—one of his past, one of his future, one of his present. The writer looks for themes in these three photographs. What complex multidimensional interrelations can he find within and between these photographs (Pinar 1994, 25–26)?

Synthesis, the last stage, brings together the three photographs. The writer asks himself questions such as, "What is the contribution of my scholarly and professional work to my present? Are my intellectual interests biographically freeing; do they permit and encourage movement? Have I gained increased conceptual sophistication and refinement, deeper knowledge and understanding, of my chosen field of study (Pinar 1994, 27)? Pinar concludes:

> The Self is available to itself in physical form. The intellect, residing in physical form is part of the Self. The Self is not a concept the intellect has of itself. The intellectual is an appendage of the Self, a medium, like the body, through which the Self and the world are accessible to themselves.
> Mind in its place, I conceptualize the present situation.
> I am placed together.
> Synthesis. (Pinar 1994, 27).

Currere in Practice

Prior to cowriting *Toward a Poor Curriculum* in 1976, Grumet and Pinar experimented with *currere* with a group of student teachers at the University of Rochester, as mentioned earlier. The authors asked the students to specifically examine their own educational experiences—what they remembered about past teachers, how their school buildings looked and smelled, who their school friends were, what they did at lunchtime. Using the four steps of *currere,* the students analyzed their thoughts about their educational experiences and the imbedded assumptions they were making both then and now. Pinar and Grumet used the *currere* process as the prelude to other forms of analysis they wanted their student teachers to use throughout their teaching lives:

> [W]e have described teaching as actions in the world that are plotted along three co-ordinates: the subjective, the teachers' experience and assumptions; the objective, the particular people he will teach, the time and setting in which they meet; and the discipline, in our case, literature, its symbols, artefacts, methods. This description of teaching asks the student to examine teaching through the three profiles of self. . . . Self-as-object emerges in the study of subjectivity; self-as-place emerges in the study of objectivity; self-as-agent emerges in the study of the discipline, for it is within its particular forms and signs that the teacher acts. (Pinar and Grumet 1976, 74)

The ontological vocation of the human species is humanization, Pinar says, ascribing this perspective to Paulo Freire (1970) in *Pedagogy of the Oppressed* (Pinar and Grumet 1976, 74). What could be more humanizing, he continues, than to participate in the autobiographical process? Because of the importance of humanization, Pinar suggested moving the autobiographical process to the center of humanities education. Pinar and Grumet proposed, therefore, in 1976 that the University of Rochester begin a two-year program on the theory of autobiography (Pinar and Grumet 1976, 75). Nothing came of that proposal. Pinar and Grumet continued to use *currere* and other forms of autobiography in their classrooms, and Grumet introduced *currere* to the University of Rochester Theater Festival in 1976, as mentioned previously (from personal correspondence with Pinar, July 20, 2000).

Where This Study Fits into the Body of Work Surrounding *Currere*

Since 1976, eleven *currere*-related dissertations have been written, five of these originating at the University of North Carolina at Greensboro under David Purpel. Two of the eleven deal with the use of *currere* in classrooms with students. The others involve either *currere* in theory but without practice, or *currere* used solely by the author (Burke 1984; Cole 1984; Feinberg 1982; Hartslee 1999; Myers 1983; Smith 1990; Wallenstein 1980; Williamson 1987).[8] One of these, a recent dissertation (Hartslee, 1999), goes so far as to repudiate *currere* on the strength of the author's three attempts to write short regression essays about himself. He does not follow the developmental four steps, nor does he record more than three regression observations. Because of his difficulties in writing these three essays, the author concludes it would be difficult within the context of a traditional school to employ the method of *currere* with students. (The author of this particular dissertation

---

[8] The ninth dissertation and its abstract, written in 1990, University of Alberta, Canada, were unavailable.

was a community college instructor at the time of that experiment.) I personally feel that he missed the mark with *currere*. In attempting to analyze his regression as he was writing it he did not observe himself; he instead wrote about himself as he wanted to see himself. By not following the four-stage development, he missed the opportunity to look for themes in his writing that might have helped explain some of the educational choices he wrote about in his regression.

Of the two dissertations that discuss using *currere* in classrooms with students, one is written by a student in a graduate education course on the teaching of English, where the professor uses *currere* with her student teachers to explore their educational life. The author then suggests a curriculum for first-year college students in remedial writing that employs *currere* and critical pedagogy as its main structure. The author proposes this curriculum based on his own experiences with *currere* in his graduate education class, not on actually working with students himself in using *currere* (Hyder 1998).

The other paper deals with fifth grade teachers and the plannings of their curriculum. The teachers use *currere* for themselves to help analyze their teaching lives, and they find the method very effective. They discuss using the method with their students, but there is no mention of their actually using the method with their own students (Daniel 1991).

Pinar and Grumet themselves, as cited before, used *currere* in their teaching, but this study may be one of the most exhaustive recountings of the method in use. In attempting to show *currere* in practice—a praxis where theory and practice meet and meld into each other—I want in part to laud a mind that is capable of imagining such a process. The fact that William Pinar was only twenty-four years old when he first conceptualized *currere* makes his accomplishment even more wondrous.

Chapter III
Currere in My Classroom

> What a small portion of infinite and immeasurable time is allotted to each of us. It is so quickly swallowed up by eternity. How small is the clod of earth on which you crawl about. Remember all these things and consider nothing great but this: do what Nature bids you, and suffer what Life brings.
> —Marcus Aurelius, *The Meditations*

An Adaptation of *Currere*

This chapter deals with the presentation of *currere* to my ecology students. The project that I use in the implementation of *currere* is the core of the first semester's work in the Ecology class. As this book progresses, the reader will see that that piece colors every other part of the class's curriculum. The method of *currere* presented in this chapter is an adaptation of Pinar's method, manipulated to allow for a concentration on ecological experience.

In this two-semesters-long study of ecology, students are not only participating in *currere*. During the writing of the *currere* project in the first semester, students are also learning limnology, being introduced to prominent writers in the field, studying population biology and energy and material flow, and reading Rachel Carson's *Silent Spring*. The *currere* project intermingles with these other areas. I place this chapter early in the book not only because I am anxious to have the reader become involved with my students, but also because I feel that the philosophical underpinnings as to how *currere* could have such a major impact on my students will be more readily understood if the reader has some sense of the space in which this project took place.[1]

Too often issues in curriculum theory are presented without examples of practical application. In this book, I want to describe my classroom experiment with *currere* and how the students and I learned together that *currere* greatly enhanced our understanding of ecology.

James B. McDonald wrote the following in 1973:

---

[1] By "space" here, I mean not only the physical space, but the mental, emotional, and social space in which this experiment took place. Just as the student's understanding of his space is vital to the effectiveness of the EA (Environmental Autobiography) so is the understanding of space in which this EA project was undertaken.

Ecology is an emerging social concern that has a corollary in the centering process.[2] It takes a unitary view of the world. Thus, the inner unity of the centering process has an outer reality in the concern for a unitary world built upon an understanding of ecology. It appears that any sane attempt to educate the young must deal substantively with the impact of man and technology on his own living environment, and there appears to be little hope that we can simply solve our ecological problems with the next generation of technological developments. Ecological problem solutions call for the same value search and commitment growing from the inner knowledge of what we are and what we can be. There is a need to transcend the linear and technical problem-solving approaches of the past if we are to survive our ecological crises. Thus, a global view of the interrelationships of human structures and activities must be a central aspect of any curriculum which purports to have a transcendent developmental view. (McDonald in Pinar 1974b, 108)

That is what I am trying to do today.

The Introduction of the Environmental Autobiography Project

The core of my ecology curriculum is what I call the Environmental Autobiography, or EA. The Environmental Autobiography is a semester-long project divided into four stages: regression, progression, analysis, and synthesis. These are the four stages of *currere,* set down by Pinar in *Toward a Poor Curriculum* (1976). I present the project in very early October with an initial classroom dialogue around two quotes: Paulo Freire's "Conscientization is the process whereby a person's false consciousness becomes transcended through education" (Freire 1996, 182) and Madeleine Grumet's "Knowledge of the world requires knowledge of self-as-knower-of-the-world" (Pinar and Grumet 1976, 38). I ask the students what they think those two quotes mean, if the quotes have any relation to each other, and what is meant by "knowledge." It is a clumsy discussion because the students are unfamiliar with this kind of vocabulary, and the idea that knowledge can be more than knowing that George Washington was our first president and seven times nine is sixty-three discombobulates them; they haven't really thought about the *idea* of knowledge. I then ask them what knowledge they have about themselves. There is quiet. Someone may say they know they are good at football and crazy about pizza, but answers stay superficial. That is when I begin to tell them how this EA project will work. "Over the next four months," I tell them, "you will be writing an autobiography but not the ordinary autobiography. This one will concentrate on your relationship with the environment. Through the writing of this autobiography with this

---

[2] McDonald is talking here about the aim of education being *centering* the person in the world, a fundamental process that makes sense out of our perceptions and cognitions of reality (McDonald in Pinar 1974b, 104–05).

particular emphasis, I hope that you will begin to see your connection with the physical world in a new way." It will be an autobiographical account of how place has helped make them who they are. In a multicentered postmodern society it is important to think about place.

When I talk to the students about place, I am talking about "space plus memory." Contemporary life alienates us from thinking about our dependence on place. If we truly understand place, we begin to create community, grounding ourselves in a web of social, political, and historical relations. Political processes are generally shortsighted about determining the disposition of urban and rural land, and these issues need to be made more relevant to the general public. Geographical location makes us part of who we are. Even though postmodernists would say we are decentered and fragmented, I believe there is a wholeness to our lives that can be found in a genuine relationship with place. And if we understand how place can affect us intellectually and socially as well as economically, we can begin to formulate a philosophy of life that incorporates attitudes of placement, displacement, and place preservation.

Preparing Students for the Environmental Autobiography Project (EA)

We will spend a month on each of the four stages. The first stage is regression. I tell students: "The first phase of the autobiographical method is writing the world, the creation of a text that brings your experience with the physical world to words. You will be remembering, feeling. You will be writing about your own experiences of your physical environment. For this stage, the writing should be chronological; you may decide to change that later on, but for now stick to relating events in the order of their occurrence.

"Record, beginning with your earliest memories of the environment. Return as fully as possible to the past, to allow yourself to be there with your mother or father or siblings or relatives or friends. Picture your past environments; describe them, not only factually but how they made you feel. If you find yourself thinking about a past experience and resisting recording it, record that resistance. Record everything that happens to you while you are doing this regressing. This is slow and hard work. What you are trying to do is open yourself up, open up your unconscious. Look at what you have taken for granted.

"Do not attempt to interpret what you observe at this point.

"You may want to use photographs to help you and include them on your pages."

There is silence. After a minute or so, someone feebly raises his hand and asks the inevitable question: How long does it have to be? When I tell them they will easily have thirty pages at the minimum, they instinctively recoil. They think I have lost my mind. Never before have they written anything that long. I tell them that is not important now. What is important is not to shut the project out, to give it a try, to open themselves up to this experience and just see what happens. Everyone is willing. I give them a week to work on the first draft of their regression, telling them we will bring our work to class and share with each other what we have done. Someone asks how far back do they have to remember, and someone else says he can't remember much before age seven. Again, I tell them that is not important now. Just start remembering. They have to record chronologically but they don't have to think chronologically. If tonight, when they sit down to work on the project and they remember something that happened to them two years ago, write it down. If the next night they remember something from when they were in kindergarten, write that down, but put it before the first night's writing. I suggest they talk to their parents and grandparents about the project; ask them for memories they may have concerning them and a particular space. Come to class next week with "data."

A week passes. Periodically, a student tells me about some memory he has written about the night before, always with a big smile on his face, wanting me to know all the details. There is discussion about the project with students from other classes. These students generally think I have asked way too much from my students, especially in a school like this where the homework load is intense and students are highly pressured in academic, athletic, and social areas. The class period arrives, and the students come warily into class, an unusual behavior from them as they usually come in energetically and full of conversation. I give a short synopsis of regression and ask if anyone wants to start off with something he has written. Steven speaks up immediately. He has memories of a tree house that his father built for him and his twin brother when they were quite young, and he reads a long segment to the class about the many experiences that centered on that tree house. (It won't be for several weeks before I realize that the tree house is on a very small city lot.) The rest of the class listens intently to his reading. (By this time in the year, we have done so many discussions in class that the students have learned the "rules" for discussion; whoever is speaking speaks without interruption and deserves concentrated attention.) But this is a different kind of attention than normal. I see lightbulbs going off in several heads, heads nodding in agreement, notes being scribbled. "Yes, I remember something like that, too, but instead, it happened for me this way . . ."

The next person reads about leaf piles. He and his family built huge leaf piles every fall for years, and he recounts several experiences with them, most of them funny, but one in which he broke his arm. Again, the response is the same—oh, yes, I did that, too, here is what happened to me. The third person to read tells of different times he found bird nests, where they were, how that made him feel—sometimes happy because he could watch the development of baby birds, sometimes sad, as the eggs got destroyed. Again, more students jotting notes to remind them to write about similar situations for them. One student tells how the Baby Jessica story affected him, and everyone begins reciting where they were when they first heard that story and how they were so frightened by it. (Baby Jessica is a little girl who fell down a well and was rescued after several days of entrapment and an entire nation watching the rescue on television.) Another tells of the first time he saw *Jaws*. The ninety-minute class period speeds by. By the end, several students have still not had a chance to share what they had written, and all of a sudden want to share, want to see if anyone else had a similar experience. I agree we can continue the process the next day.

That is how the rest of the month goes. Each student, with one exception, is eager to share what he has written. By the middle of the month, I arrange to have a one-on-one work session with each student as to what needs to be done for the "final draft" of the regression stage. (I use the word "final" here only to show the last draft before we begin Stage Two, but it is not final in the sense of no longer needing revision.) By the end of the month, everyone has at least ten pages, and there are several students already up to thirty pages. As these initial weeks go by, I am reminded over and over by the students how their initial reluctance has totally dissolved. They are writing their EAs instead of doing other homework. They go home at night, eager to write down something they have thought about on the ride to or from school. And they want to share—almost every class is interrupted by some student who brings something in that day to read to the rest of the class. Not every reading is of a pleasant occurrence. The students read just as candidly about their fears concerning the environment. One student who has moved four times decides to center his memories on a particular tree at each of his homes and what happened around those trees. One time his dog was tied up to the tree and was shot by one of his neighbors. Another time the student was climbing one of the trees and fell and broke his arm.

One of the most interesting developments is the ability of students to keep going farther and farther back in their memories once the process has begun. One of the memory tricks I gave them at the beginning was to think about the views from their bedroom windows. What did they see, and when did they see it? If they have moved often, how did those views change? By the end of the first month, students are talking about being in their cribs and watching storms out their windows or seeing moonlight or watching the sun come up. I run into parents at sporting events and they tell me how they have had interesting discussions at dinner about old vacations or the first time their son went swimming or sledding. They also mention how their son's perceptions of an event differ so from how they, the parents, remember it. I will see this idea of perceptual difference over and over again throughout the project. As one student says toward the end of the third month, "You need to publish these papers because they will let people know how teenagers really think." It brings us back to our initial discussion about knowledge. Here you have perfect, practical examples of the construction of knowledge.

Preparation for Stage Two of the EA

By the end of this first month, as we get ready to move into progression, Stage Two, I realize the project has developed a mind of its own; the students want to spend more and more class time discussing their EAs. I am no longer worried about whether the students will do the work. I am now more concerned about keeping them "confined" to the environmental limits I set at the beginning of the project. The students themselves don't want the regression stage to end. They are not finished, and new memories keep flooding them. I tell them regression is never finished and, of course, they can keep adding to Stage One, but we also must move ahead to Stage Two.

Stage Two is progression. Where regression was a calling up, progression is an imagining. I tell them: The progressive stage is more pleasing. You meditate on what may come, on what you wish to become. You look to your future and the role the physical world will play in that future.

The difficulty that comes with Stage Two is that it requires a very different mind-set on the part of the writer. This will prove to be the hardest of the four stages to write for all of the students because, I think, they have little experience in thinking about the future. I begin by giving them a technique I gleaned from Pinar and Grumet (Pinar and Grumet 1976, 58) to get themselves into this mind-set: "Sit alone, perhaps in a slightly darkened room, in a comfortable chair. Close your eyes, place attention on your breathing. Take a few slow deep breaths—you are getting into a relaxed state. After you feel relaxed, begin to think of the future, of tomorrow, of next week, of the next few months, of the next academic year, of the next three years, and so on. Since our interest is in environmental experience, gently bring the attention back to these matters, and allow your mind to work free associatively. Record what comes. Try to figure out where your environmental interests are going, the relation between these and your private life, between these two and changing historical conditions. Perhaps you will begin to see something of the interdependent nature of your environmental interests and the historical situation. You might imagine a future, perhaps a year from now or several years from now; describe it.

"Return to the chair and this dwelling in the imagined future several times on different days over a period of several days or weeks. Such drawing out of the experiment increases the likelihood of pictures that are more reflective of more lasting anticipations."

This is meditation, meditation with a specific direction. It takes the students several tries before they start to feel they have something worth writing down. I do some meditational picturing with them in class. We go outside and sit quietly and picture ourselves in certain situations. I then ask them to talk about those pictures. They learn to observe more fully what is in their pictures—how they look, what they are wearing, what the ground looks like, who they are with, whether they feel happy or alarmed. This seems to help them when they work through progression themselves, away from class, without each other to bounce ideas off of. They are struggling with this stage. I can see it in their faces, hear it in their conversations. But it doesn't stop them. Instead, they are very curious about where it will lead them. They talk of pictures coming into their heads that they would never have imagined otherwise. I tell them to write about those pictures but also to write about how those pictures made them feel. Again, there should be no interpretation. We are still "gathering data."

Another technique I developed to help them with progression was the Life List. One day we went outside and sat under some trees in a particularly quiet part of the woods. I read the boys a few paragraphs from a magazine article I had found where people were discussing the things they wished they would have done in their lives, like climbing Mount Everest or snorkeling the Great Barrier Reef (*Outside*, December 1998). I told them how many people, as they get older, have those thoughts, somewhat akin to Thoreau's "lives of quiet desperation." One way to avoid that, I said, was to set goals, goals other than marriage, a family, and a successful career. So we started right then a Life List, a list of things we would like to say we had done before we die.

The items on the list could be anything. When I asked after the writing whether anyone wanted to read us his Life List, quite a few students responded, and it was in their readings that they gave other students ideas for their own lists. We revisited those lists three other times during the year, and for a graduation gift I gave each student a framed copy of his Life List.[3] This technique for thinking about the future helped some students with progression.

---

[3] The last time we worked on the lists, just a few days before the seniors' last day, I asked them to compile all the items they had put on their lists and make any revisions they felt necessary. It was interesting how many chose to remove some of the more superficial items, such as getting a tattoo, and emphasize more goals based on environment.

During Stage Two, the students were still intent on sharing what they had written with each other, but there was not the "lightbulb" connection of the first stage. These progressions were much more idiosyncratic. Each student had his very own specific pictures of his future self, and while interested in hearing what others were writing, he was not interested in adding their pictures to his collection.

I found my job changing somewhat in this stage. I now needed to get some of my students to stretch themselves more. Whereas the regression was flooding them with memories and they were not yet having to be selective about those memories as long as they centered on place, some students were having very few pictures in progression, and some of those pictures were quite shallow. I did not discourage those pictures—for example, pictures of them playing professional baseball, signing autographs, adoring women surrounding them—but it was not enough. But once I began this push to think more deeply, I found the students passing judgments themselves on what their peers were writing. This was not happening in the first stage. Then everything was considered important and worth writing about. Now the students seemed to feel a need to write about pictures that showed a depth of thought and perspicacity. It was the first time they were critical of each other. Even in other discussions we had had up to this point, discussions not concerned with the EA, students had listened to each other without voicing judgment (probably thinking judgmentally but keeping those opinions out of the discussion). Now it was different, and there was a foreshadowing of what was to come with Stages Three and Four.

Preparation for Stage Three of the EA

Growing

I live less now
Memories are overwhelming
Adventures were everywhere
I had time for myself
And now
I have no time for adventure

Playing in the woods
Splashing through streams
Sliding down hills
I had the best of times
Playing in the woods

As I walk in the woods
I see footprints of the past
They are smaller

They are perfect

Those footprints have been there
But new prints have been made
The new are bigger

The new prints are less perfect
When comparing the two

But as I look back everything seemed bigger.
      (John, written as part of his Stage Four)

We had now spent two months on the EA. No one was talking anymore about the number of pages expected for the assignment. Instead, there seemed to be much introversion about the future. The seniors were in the thick of college essays, and there was constant discussion about what college was "best," and who was applying where; it was not only the EA that was bringing the students to thoughts of their futures. They were anxious to begin Stage Three, analysis. I told them: "Now is the time for critical reflection. Reading your autobiography up to this point, you begin to look for patterns, for themes. You will try to link the present with the past, the past with the future, the present with the future. What areas are you drawn to? What areas repel you? You now need to detach yourself from experience and try to answer the question, why are these experiences as they are? You interpret.

"This stage is different for everyone. It is impossible for me to give you a 'recipe' for it."

It was December, and we had a little less than three weeks for this stage. I worried that this critical stage would be rushed and the upcoming holiday break would be distracting—as if we hadn't already had plenty of distraction with the college application process. But the project again brought the group to focus, and the intensity switched to a different kind of abstraction from where we were with progression, Stage Two. I suggested that they read and reread what they had written so far and try to categorize different experiences—those that were fearful, those that brought surprising results, solitary experiences, and so on. What made a situation fearful? Would you still be fearful if confronted with that same situation today? Have you had many experiences, and do you foresee yourself having experiences outdoors by yourself? Do you always see yourself, whether in the past or the future, in the midst of a large group of people? Questions such as these helped them start the analysis process. An eighteen-year-old is not always capable of deep discovery. I was hoping that in class discussion we could help each other pursue some of these analyses in a more substantial way.

But once again, I was surprised at the turn of events. Once this stage was begun, almost no one wanted to read anything he had written out loud. Now, for the first time, I heard conversation about the need for privacy with the EAs. There was really only one student who continued throughout the process to want to share everything he had written, and I believe that had more to do with the total lack of duplicity in this individual. He also is funny and enjoys making the class laugh, but he is insightful and thoughtful and produced an excellent project (more about this individual later). Everyone else closed down, and it was at this stage that I learned some startling facts about some of my students.

There was a sense of catharsis for many, and an admission of feelings they, or at least as they perceive the culture, would consider unmanly. One student admitted in class that he stood up from his desk at home at one point and told his brother he was being too sensitive, he was becoming a "wimp," yet he couldn't help himself, he had to keep going. When he related this episode in class, heads started nodding—yes, they had experienced something similar. Was there something wrong with them? It was a measure of the synergy of the class that a discussion like this could take place without worry that it would be repeated outside the classroom doors. I asked them if they could honestly tell me how it felt to be "sensitive." One of the biggest jocks at school spoke up immediately. "I liked it," he said. "I noticed that for the first time today I watched the sun coming up over the driveway into school and said quietly, to myself, how beautiful it was and how lucky I am to be able to see a sight like that almost every morning." Right away, everyone was agreeing with him, most needing that first fellow to come forward and say what they were feeling. I continued to recede into the background. Within the security of the classroom, the boys talked freely about becoming sensitized to their feelings but not about their own individual analysis of their project. That was too personal for them; they were too unsure of how the group would react to their exposure. The bell rang and we all just sat there looking at each other. No one wanted to end the discussion, yet in some relieved way, each felt glad to have escaped detection. They passed me their papers quietly on the way out.

I used the holiday break to write them long notes on their analyses and give suggestions for further development. When we returned in January, we would have less than four weeks to wind up the project and prepare for exams.

Preparation for Stage Four of the EA

Stage Four is synthesis. Now comes the integration. The students were back, rested from two weeks of break, and all college essays and applications should have been finished by now. I told them, "This final stage is integration. You are achieving an understanding of yourself and your relationship with space that is both intellectual and emotional. This understanding gives you freedom to act. You begin to think about yourself in a different way. Record these changes in yourself; record the changes that this whole process has brought about in your thinking about yourself. What have you discovered about yourself that surprises you? How does this process make you think about the way you want to live your life, both your personal life and your public life? Your physical body connected to your physical environment?"

The writing attitude was back to the way it was in Stage One. The students wrote easily and copiously. They had much to say about the project itself and how the project had affected them. They wanted to write about how differently they saw themselves, how they had learned why they have certain fears and certain experiences that they love unqualifiedly. In class they were willing to talk about the project per se but not about their personal involvement with the project. Discussion narrowed and there was again much talk about the need for privacy with the EAs. Students from other classes came to ask me if it was indeed true that Student A had written over 100 pages and could they please see them, or that Student B had actually typed fifty pages himself—they didn't believe it. I was very careful to guard the work and maintain the students' privacy.

As exams approached I reminded the students that we had to end the EA. If I was to get my grades done according to the school timetable, I had to have time for a final reading of each student's project. There was much discussion about whether we really needed to end the project now, and even though secretly I agreed with them, I had another project I wanted to try with them second semester, and I insisted that it had to end now. I did promise we would revisit the EAs in the spring, which we did. I collected the papers, and, for now, the EA ended.

Summary

I had hoped the EA would begin to emotionally connect the students to the environment. During the times we were exploring the basic scientific principles of ecology, we were also exploring the interior lives of people interested in ecology—themselves. Even though I felt the method of *currere* was excellent and theoretically sound, I was unsure how using it in this context and with this focus on place would work with eighteen-year-old males. If it turned into an assignment like any other at school, where concern with the number of pages and meeting the deadline took precedence, it was not what I was looking for. I needed to find something that would move my students from "I know" to "I care." I believe I did. The EA engaged the students and brought an enthusiasm and energy to the class that I would never have predicted.

Chapter IV
Data from the First Year's EAs: Discovering "Nature's Temporary Cushion"

> That man is, in fact, only a member of a biotic team is shown by an ecological interpretation of history. Many historical events, hitherto explained solely in terms of human enterprise, were actually biotic interactions between people and land.
> —Aldo Leopold, *Sand County Almanac,* 1949

The data presented in this chapter represent sections of EAs from the first group of nineteen students I had in this experimental ecology course. I have chosen samples from each student to try and show the idiosyncratic approaches taken to the project. I am attempting to show something unique with each student and have left no student out.[1] For most students, I begin with a selection from his synthesis so the reader can see what that student felt about the project as a whole. The other choices for that student have to do with my desire to show something specific about him and how he used the EA. The order in which the students are presented is strictly random.

There is no denying the instability of memory, whether the person remembering is eighteen years old or eighty. Memory is constructed, as is any other reality. Kathleen Weiler (1995) writes about memory as having two distinct parts: the material reality of the remembered experience, and popular memory, where the past is revealed not as a set of facts, uncovered through the interrogation of eyewitnesses, but "as a social construct expressing power conflicts and competing meanings" (129). My students' memories cannot escape the hegemonic atmosphere in which those memories were created and remembered. I attempt to point that up in some of these excerpts; in other excerpts, I let the student's writing speak for itself.

Excerpts from Students' EAs

From John's synthesis:

> As you could tell from the writings of childhood memories, I've always liked being outside. If I wasn't outside then I was thinking about being outside. I can al-

---

[1] I cite from each student's work for two reasons: first, to show the myriad ways students used and benefited from *currere*; and second, to ensure that I could not be accused of having an agenda to prove *currere*'s effectiveness by quoting only those students whose work was deemed by me to be "top-notch."

ways remember two things about coming home from kindergarten, one was thinking about playing in the stream in the backyard, and the other thing that I will always remember is showing my mom the art work I had done that day. It's strange that I pictured myself in New York City in my future. I think that deep down I just want to capture the generic American dream of working in the city and having a nice wife who wears an apron. [I commented heavily on THAT sentence!] It must be imbedded in my head somewhere that this is what society and my family would like me to do. Even though I would much rather be a guy who has a nice tree in his yard so he and his son could climb up into it and tell fish stories. It just seems that if I'm going to dream I might as well go crazy. When you start thinking about your future your mind kind of pops out neat things that you like so you make them your future. When I was thinking about my future my train of thought went like this: I like art, I like to go to the city, I like penthouse apartments, I'd like to be married, and I like cars. Hence leading to my future through immature eyes. I think that I am an immature person that really relies on my parents, I don't worry about college but I give signs that I don't want to go, like not filling out applications, and avoiding Mr. Lloyd [our head college guidance person]. My future is only a dream away. I am so carefree that I don't no [sic] where the wind will blow me to, maybe NYC, maybe Nebraska. In my opinion, when it comes to dreaming and wishing you should go beyond your goals. For example, if your [sic] stranded in the desert and you wish for "just a drop of water" why don't you just wish for a pitcher of water with ice cubes and maybe some ham or a chicken sandwich with mayonnaise? It's a good question.

More from John's synthesis:

Throughout this project, and up to the end of it, I really took the time to think about myself. I always think about stuff but it really felt good to give some thought to me. The project brought out true feelings and thoughts that I am grateful to have written down so I can read it in the future. This journal is an exact equivalent of a photo album. I'll look back on it and really thank you, Mrs. Doerr, for not making me, but requesting me to do this literary photo album.

John was not someone anyone ever told what to do. I am not sure how he perceived the project as somewhat optional, but if he did, that is good because he bristled at the thought of having to do something just because someone in authority said so. John was an artist and a free spirit, very gifted with pencil and computer graphics. John had grown up with a family that traveled widely; he lived in Austria for awhile, and he had many wonderful stories to relate in his regression. He was very family oriented and struggled mightily with the thought of going away to college and leaving his parents.

On the surface he appeared very together, warm and friendly. He drove an old police car and had a policeman's shirt, hat, and badge that he wore on occasion to school. He loved car mechanics and worked an after-school job at a local garage. He was using his EA not only to make his "literary photo album" but also as a kind of catharsis, I think. For example, he wrote in his progression a story of his future that belied his apparent self-confidence:

After I graduate from college, my name is known all over and people are wanting to hire me everywhere. Major advertising firms in New York City are recruiting me. This world class place hires me and puts me in charge of magazine ads. I get to design advertisement layouts for companies like Microsoft and Milk. I already have enough money to afford a penthouse apartment in downtown NYC. My wife and I move in and my grandma, as she promised when I got my first apartment, gave me her tan leather sofas. The wife and I put the couches in the family room and sleep on them until we by [sic] a bed. I am living just how I wanted to when I was in high school. I drive a black 2006 Ford Crown Victoria police edition with the interior customized by me. My car is just a toy because I rarely ever drive it, I usually take the subway or taxis. It feels good to be loved, not only by your wife and family but the people you work with. I'm so young but have such a good start on my future. I am really proud of myself, throughout high school I was always the "dumb" one but look at me now, shining like a star in the sky, and on top of all, I am happy and loved.

John had an older brother who was suicidal and had had several bouts of severe depression, causing him to leave college for awhile, then go back, then leave again. When his brother would come home, it was expected that John would spend considerable time with him, the thinking being that John's natural good humor would help elevate the boy's depression. It worked the other way around, though, and John had difficulty keeping himself buoyed up. Most of the time, though, his thinking was very positive. He would often have the most wonderful, unusual take on things, that artist perspective.

From John's regression:

Just a couple of hundred yards away from my house there is a little stream. There is a small waterfall that makes a really pretty noise when it trickles into the pool beneath. This is where I go to be with me. I think about things that most people don't even consider thinking about. I wonder things like if everything on our bodies is an even number (i.e., two eyes, legs, arms, nostrils, ears, feet, hands, and so on) then how come we don't have four fingers on each hand. You may say that you have one tongue or one nose. But I say you have two halves. So in my opinion, toes and fingers should only add up to sixteen.

More regression:

Whenever I have a campfire I notice how much wind there is. If there is none I wonder why the smoke always is near my head. I pondered and pondered, and the only reason I could think of might sound stupid. My mom always told me to wear a hat in the winter to trap the heat my head gave off. I thought that since my head gives off heat that it is making a little river, or wind, in the air. The smoke gets sucked into it therefore causing the smoke to constantly follow my head. If you know the real reason, please tell me.[2]

More regression:

Whether I was at home or someone else's home, I always had a fort, a lair where we could hang out. One of the forts I made was with my friend Jeroen. He and I dug a hole in the ground about 4 or 5 feet deep in the middle of the woods. I was about 14 years old. After we were finished with our three-day-long expedition, Jeroen and I laid sturdy sticks over the hole, and then put branches and leaves over that. It was really neat because no one would ever be able to tell where it was even if you were looking for it. There is something enchanting about a fort. Jeroen and I never spent too much time in our fort, but it was still a good place to hang out. The most attractive thing to me about forts, hidden forts in particular, is that it is very reassuring knowing that there is a place you can go where no one will ever find you. The mystery of being hidden has always intrigued me.

John had good memory for detail and all his regression episodes reminded me of old-fashioned summers and boys spending hours outside entertaining themselves away from TV and too many programmed activities:

[John and his brother are out on a sandbar with their father diving for shells. John is eleven years old.] On my fourth try I caught a big conch shell. It was about 8 inches tall, and it had something that looked like a tongue inside of it. I asked my dad what it was and he told me it was the muscle inside, just like in a clam. As the day went on we caught 70 of these conch shells, 20 sand dollars and 10 starfish. We put them on our air mattress and brought them to shore to show my mom. She took a picture and then my dad, my brother and I took them back to the sand bar and put them back. We kept 5 sand dollars, and soaked them in bleach, so they'd be clean and white. My grandma and grandpa still have one of these sand dollars hanging above their kitchen sink.

[Picturing Tom and Huck]: Every time I went to Mike's my mom said, "Don't swim in the river" and every time, I did. We would swim from the big green bridge in Gates Mills all the way to where it met Mayfield Road. We would explore all the little streams that let out into the Chagrin. Most of the time, there would be nothing but bugs and pricker bushes, but one time we found the mother spot to hang out. It was a healthy running stream about 6 feet wide in spots. It was about 800 feet from the Chagrin, and it was completely surrounded by trees. This spot had an eight foot deep swimming hole, surrounded by clay, soft clay that you could make stuff with and at the bottom of the swimming hole there was sand and small pebbles, no sharp rocks. Oh yeah, there was a ten foot tall pipe that you could jump off into the swim hole. This place was awesome. It was so secluded. We used to bring lunches there and lawn chairs and towels, just like going to the pool, without the lifeguard telling you to stop splashing, saying you can't jump off that, just me, Mike, lunch, and nature.

---

[2] Of course, I notice the same thing when I am at a campfire! Will I ever be able to be at a campfire again without thinking of John and picturing the "river" of wind over his head?!

From Gene's synthesis:

I honestly can say that throughout this autobiography I have actually wanted to sit down and type for hours. [This is from a student who has been on and off ritalin for several years, diagnosed as ADD, and who cannot sit still in a desk for the average 45-minute class.] When I type this paper I don't worry about making grammatical errors or spelling mistakes because what really matters is the contents of the paper. I am sure that when someone writes an autobiography he or she does not focus on a perfectly structured paper but he or she focus[es] on what message they are trying to deliver. When I type other papers it will take me hours to think of one paragraph. In that time I could write two to three pages about myself because the words just flow out.

More from Gene's synthesis:

First off, I would like that say that in the beginning I was a little skeptical of how this whole paper would develop and in the end how it would turn out. Unlike the McKibben article [a long essay we had read in September and analyzed critically, an article in which Gene could find nothing he agreed with] I decided to give this autobiography a second evaluation. I must say that I really have enjoyed writing this paper and I have learned some things about myself that I was not aware of. Usually when I have to write a paper it is considered a hassle and I don't want to take the time to sit down and think about it. With this autobiography I felt like I could actually write whatever I wanted and I could also talk about how I was feeling. I think that this was a good experience altogether because I feel that this paper will not only open up some doors that some people have never really seen inside before, but I think that it has at the same time opened me up. Although some of the things that I have talked about I would not like some people to read because I feel that they would criticize me, and I really don't want to deal with that. I suppose the reason that I can write what I am thinking and how I feel is because of my teacher Mrs. Doerr. I do not really know why but I feel that I can be more open about things if it is a woman that is reading my writings. Maybe it is because some guys are always trying to be so macho, and they never really look inside themselves and see what it is that is making them feel this way. I would be a hypocrite if I said that I have always been this open because I have not been. The main reason that I am is because of my past experiences with my girlfriend. She has taught me to be myself all the time instead of putting on an act just to appear a certain way. Now that I have almost been with her for a year I must say no matter where I am, I am myself. I figure that if someone can not accept me for who I really am, they aren't worth the time of day. . . . Once I got the hang of it, expressing how I feel is very easy, and I think that it has made my writings better.

Gene was a fellow who loved the outdoors. He had not originally signed up for this course—he was going to take AP Biology instead, but two weeks into the school year, he asked to change to this class; AP Bio wasn't quite what he had in mind. He was a three-sport athlete, captain of both the hockey and baseball teams, and did not want to spend hours every night working on vocabulary. I could see he was somewhat reluctant to take this class but agreed to take him on. It was one of the best decisions I made all year. His love for the outdoors was so deep that it transferred to everyone in his class. His enthusiasm was palpable. With very rare exception, he was intensely involved with everything we did.

Gene's analysis:

> I sometimes wonder if there is a special chemical in the air that magically reels me in every time I step outside. In order to let you know exactly how I feel, I think it is necessary to fill you in on the certain aspects of each season that make me feel as if I am in another world. [Gene has taken this "talking to the reader" tack throughout his EA.] Let's start with fall. When the trees are changing color and the leaves are slowly falling, I feel as if I am walking in another world. I feel as if everything is painted perfectly on canvas and everywhere I go someone is quickly putting up the different sets. For instance, when I am driving up [our school's] drive and all the trees are sporting their favorite colors, I am almost swept away completely. Those colors make me unfortunately want to be anywhere but school. Since I know that I have no choice I let the trees put me into an extremely good mood, and I slowly make it through the day. When the first real chilly evening comes, I get excited. I get excited for several reasons. The first thing being I can finally where [sic] all of my favorite clothes in order to keep warm. The second being that I can finally make a fire in my house and the third being I can go to the park with my girl friend and lazily walk through the woods where all the colors put a mystical feeling in the air. The next season, as you know, is winter. Winter does wonders for me. I love the heat, but I love the cold just as much, if not more. I feel completely at peace on a cold, wintery day. For some reason, winter carries me off to another place that is unmatched by any season. Perhaps it is the Christmas spirit that is in the air and all the lights that are found on most houses. Or it could be the Christmas trees. This year I got my mother to buy two just so I could be in either family room and enjoy the beautiful sights that each tree brings. I love putting on the snow suit and running wild outside and then coming in to warm my bones by the fire. I don't know why but I actually like being cold. I love the morning drive to school. I only turn on the heat for a little, just enough to defrost my windows, then I put on my gloves and hat and enjoy the ride. The other things I love involve going downtown to Tower City and shopping for Christmas. I must say that once Christmas is over and the winter is coming to an end, I feel sad and I kind of panic. It is just like when you are going to lose something and you can't do anything about it. I hate knowing that I have to wait another year to experience the joy of winter again. But as always, I slowly get excited for spring.

He went on from there to talk of both spring and summer in the same poetic way. This macho-looking, publicly macho-acting boy was discovering his soul. In class, when we were outside performing an experiment or writing an observation, he was a different person than the one I would see in the hallways or the sports fields. While he was still almost hyperactive, his energy was routed in a very positive track. He wrote beautiful observations in his journal, often while sitting in the top branches of some tree, while the rest of us were sitting on the ground below. I wrote copious responses to his observations and he almost always commented on what I had written back to him. He liked living "on the edge" and was the one who would test the ice for us to make sure it was safe to cross the lake or cross a muddy bog to attempt to capture a vole. In early October he brought a ball python to class and asked if we could keep it as a class pet. For months the python "sat" with Gene during each class, enjoying the warmth of his hands and body.

From Gene's progression:

> When I started to write this section I did not really think that I would be able to think of some occupation that I could really see myself doing. Now as I sit here and write, the possibilities seem endless. I think what I really need to do is choose a path that would definitely keep all these opportunities open to me. . . . I definitely have to avoid the business man trap and find out exactly what I want to do. I know one thing for sure and that is that nature is definitely one of my loves, and I think it is only right that I pursue it. Where I will end up I do not know, but hopefully I will not end up in some uncomfortable chair looking out the window realizing that I missed the opportunity to do something that I love.

One of the advantages of Stage Two is just what Gene has written about here—to be able to think about options for yourself in the privacy of your own mind without worry that you might disappoint your parents or offend some relative. Especially in a school such as ours, the idea of a career in environmental studies is considered "odd." Success is equated with monetary reward. It will be interesting to see how committed he stays to such a dream for himself. He has already chosen a college with a strong environmental program and has signed up for the environmental studies track. I doubt that this would have happened if he hadn't gone through the EA project.[3]

Craig had chosen to title his EA something other than "EA": "In to the Mind of Craig." He wrote me a letter after graduation in which he thanked me for giving him the opportunity to "enter his mind" and explore some of his early behaviors. In his EA he wrote in great detail about his relationships with space, but in the end he used the EA as catharsis for something he did in his early teens that he was barely beginning to understand about himself.

He began his regression:

---

[3] By the end of Gene's freshman year in college, he was well on his way to majoring in environmental studies. During his sophomore year, however, he became disillusioned with school for various reasons, and dropped out of college for a year. He is presently matriculating at a different university.

When trying to remember your life, there are certain things that stick out. Smells, feelings, colors, temperature—these are all things that are fairly easy to remember, but it is hard to remember when things happened. I found that I can usually remember where they happened so I separated my memories into five categories. I have lived six different places in my lifetime, but I only have memories of five of them. I found that I can associate most of my memories with where I was living at that time, giving me some chronological order. Once I separated my memories in this respect my Mom, Dad and brothers helped find order within the categories.

Today Craig lives right on Lake Erie. He swims in the lake daily, missing only the coldest part of the winter. Water is a major theme in his EA:

My life to my recollection begins in the summer of 1983. I was 2 3/4 years of age. We lived in Conneaut, Ohio, a small town that is situated on Lake Erie. The town's main industry involved the loading and unloading of ore onto or off of ore boats. It had one of the fastest facilities in the world. As a little guy, I remember how these huge boats fascinated me. My grandparents lived down the street from me and they had a boat. When we lived in Conneaut we went out in the boat almost every other day all summer long. When we took boat rides I always wanted to go see the ore boats. They would pull into this channel where the boats would park until their purpose was accomplished. There was enough room for a small boat, such as my grandpa's, to ride right alongside them in the channel. You have no idea how large these boats are until you're right next to them. My first recollections are of these large boats and my father always being afraid as my grandpa drove near them. I was anything but afraid. I loved those boats, I wanted to get close enough to touch them. It was so hot when we were near them, and it smelled funny because of the gasoline fumes. The water in that channel was also very dirty. You couldn't pay me to swim in that water.

That same summer, once the water became warmer, my mother would try to teach me how to swim. This was not a problem because I so badly wanted to swim. I loved water then and still do. We would go out in the boat once my mom got home from work and we would park and anchor in a sand bar. Here the water was about three or four feet deep. It was over my head, but it was comforting to see my mom could stand there. She would hold my underside while I kicked and paddled with my arms. One time she moved her arms and backed away, and there I was swimming after her. I cried because she let me go. I learned to swim, however, and have been doing it since.

From Craig's regression:

I've never been a big fan of winter. My first recollection of winter is not a good one. There was a lot of snow and it built up on top of our house. Finally a good amount of it collapsed in front of our door and just missed my grandpa. For a long time as a little kid I was afraid of the snow because of this. Winter made everything I loved about the summer impossible. The lake froze that winter, and although now I think it looks kind of cool when the lake is frozen, back then I wasn't so fond of it. I loved to go swimming but those huge blocks of ice and the cold air made it impossible and it made the air around me depressing.

He has another winter story, which at first he doesn't connect with his aversion for the cold, but as he and I worked through his analysis section, he realized this story in large part colored his feelings about winter:

> There are some memories that are so strong that I feel as though I relive them when I think about them. Christmas time as a five year old was one of them. I was sitting with my brothers in our blue TV room. . . . It was very cold in our house that night. I remember I was holding a green and white blanket that I always used to carry around with me, when my mother answered the phone. I don't know who it was that told us, but my mother's sister had passed away. My mother cried and my grandparents were shocked. They all told me she had died of a blood clot. That winter was followed by each of my aunt's children staying with us. Everybody was always fighting. I remember one particular day when my grandma told my cousin that she had caused her mother's death. I remember thinking how could this be. How could my cousin cause her mother's blood clot? My cousin ran out into the cold winter day. I went with her. She was about 17, I think and she was crying. I rarely saw old people cry. Six years later when telling my mother about what my grandmother had said to my cousin, I came to the horrifying realization that my aunt committed suicide. I couldn't imagine what it would feel like to have my own grandmother tell me my mother committed suicide because of me. It is the worst thing I've ever heard anyone say to anybody, and I pray I never hear anything like it again.

This aversion to winter played out in Craig's choice for college. From the beginning of the college selection process he had wanted to go to California, southern California. He had visited there when he was thirteen and loved it there because "it was as if winter didn't exist." He wrote in his progression:

> The one thing I will miss in California is my lake. There's a slightly larger body of water out there, but it doesn't have the sentimental value that this one does. I have stood on the beaches out there and it's beautiful, but it's not home. I often think that it is little old Lake Erie that will bring me back here . . . in the summer.

In Craig's analysis, he realized his connections between winter and sad things that had happened to him (his parents broke up temporarily in the winter, too):

> As I dig deep in my mind, I began to wonder if the reason I want to go to California is to get away from some of these bad memories, or at least prevent new ones. I know this won't be the case, but maybe subconsciously that is what I'm feeling.

He continued his water theme as he wound up his Analysis section:

> The symbol for Taoism is flowing water. Taoists believe that in the Golden Age Man was in complete harmony with nature and that his ego drove him away from it. They believe man still has it in him to understand nature, but his ego again gets

in the way. As I stare out into the lake I see pieces of myself. The lake is what links me to nature and to my past. The flowing water of Lake Erie has put me to sleep many a night, and the lake is tied into many of my memories. It is a constant in my life. It is bigger than any of my problems. We constantly discuss in ecology class how man should try to understand nature more before he attempts to tamper with it. From writing this biography, I think man would understand nature a little better, if he tried to understand himself. Everything has an effect on everything else. Each thing that has happened in my life has made an impression on me. If any thing I've written in this paper hadn't occurred I'd be different, just as when I throw a rock into the lake it moves a little bit different.

From Wayne's synthesis:

> This project has taught me a lot of things. At the same time I have learned a lot about myself. I never really liked to talk about my life. It is a lot easier to write about your life than it is to actually talk about your life. I don't really know how I want to live the rest of my life. I mean, well yeah I want to graduate from college, and be successful in whatever it is that I do. What it is that I am going to do, I have no idea. I have set three specific goals for myself; the first one is to be kind to every person I meet. The second is that, like my father did for me by sending me to a private school; I will work my hardest to make sure that my kids have that opportunity. In order to accomplish my second goal, I will have to fulfill my third goal and that is make sure I work as hard as I can at whatever it is I end up doing with my life.

Wayne, probably more than any other student in either group, used the EA as catharsis. He had been shuttled back and forth between his mother and father for many years, and both his parents had struggled through their divorce. The father had remarried; Wayne had been spending more and more time at his father's house, and he was not getting along well with his stepmother. Throughout the EA, he used the space of both parents' houses as more than backdrops for his behavior; he almost anthropomorphized the space, those places being very different and producing very different feelings in him.

Wayne was also one of the more concrete thinkers of the group and had difficulty with Stage Three. He did, though, attempt to make some connection with place as he pondered how things might have been different:

> Sometimes it seems that everyone is against me. I know their [sic] not, but that is the way it seems. This makes me question myself. What if I was somebody else? What if I had lived with my father instead of my mother? Would I still resemble my mother as much as I do? I think I would be a totally different person. It is way too confusing. That is one reason I don't like to talk about it. The other is simple because, no one will understand me as much as I do. My parents used to send me to a psychologist; I hated it more than anything. Why? Because they are trying to

analyze me. Trying to put me into a category along with every other human. I don't fit into a category. I am who I am, and I know that. I also know that there is no one else on the planet that thinks and feels the same way I do.

It took Wayne a long time to work through his regression. For the first several weeks, he wrote nothing. He said he couldn't remember anything because his life wasn't like anyone else's in the class; all of their memories had nothing to do with him. He didn't know where to start. Once he did, however, he wrote and wrote, mainly about traveling back and forth from house to house trying to decide what was expected of him at each place. I let him start Stage Two later than everyone else because he was still in the throes of regression at that point. However, that may have been a mistake on my part, because when he was ready for progression the group was well into theirs, and he never did get the idea of constructing pictures for himself. He ended up fighting the whole idea of progression:

> This exercise does not really help me picture my future. What is the point of planning out your future when you could conceivably be dead the next day? Then everything you just planned out has no meaning because you are dead. Also, one of my classmates was right when he said, that nobody wants to talk about the evil side of themselves. When you look into the future you only think about what you want, not what may happen. Everybody assumes that he or she will be a good person.

From Nick's analysis:

> One thing that I am truly grateful for is the fact that through my parents and my grandparents I have received an appreciation of nature. On my grandpa's farm and at his house in Chautauqua he has pointed out many really cool things that I would never have seen without his help. My parents help me to appreciate nature but in different ways. Looking back I can't count how many times my parents have made me turn off the TV and go outside or go to the metroparks. Through these subtle ways my parents have made me aware of my surroundings.

Nick was probably, although one of our youngest members, the most practically gifted in the group. Whenever we were doing an experiment and something went wrong, wasn't working correctly, or wasn't working at all, we turned to Nick to fix it. He was clever and creative and could come up with solutions quickly. He enjoyed having that role in the class, especially since he didn't consider himself a particularly good student. His handwriting was almost unreadable, and he had little facility with computers, so in most of his classes his work output was minimal and often discarded because of its appearance. Because of this he had created a spiral for himself of poor or non-production and a procrastination sometimes bordering on the ridiculous with assignments. But he really got into the EA and I learned to "read" his handwriting.

From Frank's synthesis:

> This whole project has made me realize that the way I act and the things I do are all because of the way I was raised. The things my parents allowed me to do and didn't allow me to do have influenced me to be the type of person I am. . . . Although I have discovered some very important things through this process, I haven't discovered anything that surprises me. Everything I wrote about I knew was already there simply because I have thought about my life in the past, present and future before. I had also already thought about what type of person I am and why I am that particular way. How I portray myself daily was also something that I had given thought to before we started this project. But this is sort of what this whole process has helped me to find out. I found that thinking ahead and wondering about why things happen is a part of my personality. My personality is why I know almost exactly how I want my life to be in the future. It is also why I know how the freedom my mother allowed me to have is why I feel secure while doing things by myself and away from home.
>
> . . . Although I haven't been surprised by my discoveries, this was a great project that I enjoyed doing. This was the perfect project for me. I love to analyze things. This process has not only given me the chance to do that, but it has also given me the chance to write down the results of my analysis in an organized manner. This probably is something that I have been wanting to do for a while but never had the initiative to do it. I am glad I got all of my thoughts on paper because it is good to know what makes you act what way. Everyone should do this project.

Frank was one of only two black students in the ecology classes the first year. He had spent his whole life in the city, living with two parents who worked many hours to pay the tuition at this school. He was a quiet boy. The rare times he would speak up in discussion or volunteer to read some of his work, the group would become extra quiet so they could hear Frank's soft voice. His regression stage memories were often very different than those of most of the class:

> One of my earliest memories is one of happiness. I was about three or four and I had just got a new Big Wheel. I used to ride up and down the street all day long. I was only allowed to go to the house next to me where my best friend lived. I would ride over there and we would play Miami Vice around the house. I also remember wanting to ride through the leaves so I could make the leaves go everywhere when I stopped really fast. Because of this, I did great damage to the plastic tire that wasn't made to handle skids. It only lasted about two summers. But I rode

it until it almost fell to pieces. Around this same time my family took a trip to Atlanta, Georgia to visit my grandmother. We would visit her because she adopted my mother and that was the only connection we had to my mother's side of the family. As time went on, we would visit more often because she became blind and very ill. I was too young to know that stuff then but I was always happy to go because I loved to get out of the house. I can remember leaving about midnight because my father likes to drive at night when no one is on the highway. I sat in the front of my father's Cadillac between my mother, who was on the passenger side, and my father, who was driving. My two older sisters and one older brother were all sitting in the back. It was pretty crowded but I didn't mind. I kind of liked it. I loved to take trips. I would try my best to stay up the whole trip so that I could see the trees and beautiful night sky. I also always hoped to catch a glimpse of some cows or any other animal that I didn't see on a regular basis. I could watch us go from the inner city to the country in what seemed like only a half mile. And if it rained, it was even better. Looking back, I also tried to stay awake for about fourteen hours because my father did and I wanted to be just like him. He seemed like a hero to me. He knew the way to everything and never got lost. He also was able to take our family places people around me didn't often have a chance to go. Although the majority of this trip was fun, I can recall a tragedy that I remember very clearly. It starts at the playground that was across the street from my grandmother's house. I was spinning rapidly on the merry-go-round due to the excessive speed that my brother and sister used to push the seemingly harmless ride. Suddenly, I lost my balance and fell. My face hit the gravel. I busted my lip and blood was everywhere. I didn't go to the hospital even though it sounds like I should have. I just put a cold wet rag on it that had ice in it. A big lip was not the only thing that I remember taking home from Atlanta though. I also recollect going outside at nighttime with my father and getting bit by huge bugs, which, I would later learn, were mosquitoes. They would leave big lumps on my legs that would itch for days but I still loved being in that sort of forestry environment, and being with my dad made it twice the fun.

Frank always loved going out in our forestry environment at school. He was a sharp dresser, pants perfectly pressed, shirt beautifully tucked in. Somehow, whenever we were outside, even though he participated in everything, he would return to the classroom looking as sharp as ever while the rest of us looked like we had picked up too much mud and sweat.

Frank really got the idea of Stage Three. He decided to use themes to organize his material, themes centering around different aspects of his personality; for example, shyness, self-sufficiency, independence, trustworthiness. He linked episodes in his regression and progression to these themes and built up a scenario for himself in which these themes connected with his present space. He then linked these perceived personal characteristics to their presence or absence in his father. An example:

> I just really don't like depending on anyone. This is why I work so much and so hard. I don't have to ask my parents for anything. All of this came from my father. I remember he told a story about when he was little and still lived in Alabama. He said he was starving but there wasn't any food in the house. His mother told him

to go down the street to a friend's house and ask to borrow some potatoes. He walked down the street and pretended as if he was going to the house that his mother had told him to go to. He walked around the corner and came back without the potatoes. He told his mother that they didn't have any even though he never really went to the house. He said he'd rather be hungry than ask some one to borrow something. Some people see this as foolish pride but I know I would have done the same thing. A week later my father got a job and started buying food himself. I would have done the same thing.

Frank was very proud of his work with the EA. He would often spend a few minutes with me after class checking something out that he wanted to add. The last week of school he decided he needed to do something that was out of character for him; he told me his life was too predictable. So he went out and bought himself a used car, a Chevy Impala.

From Sam's synthesis:

> As I have been writing this paper I think that I have become even more aware of my surroundings. I am constantly looking for new things to write about and new things to experience. I am more open to taking a walk through the woods. Before I began writing I would have only gone if I was asked to go. Now I thoroughly enjoy strolling through the woods on a nice cool crisp winter day or on a beautiful fall day. Not to say I did not enjoy the environment before I wrote this but I did not make the effort I make now to expose myself. For example, I would be at football practice, somewhere I had to be and I would notice the environment. I remember saying to myself where else would I want to be at this time. But now you may see me playing in the snow-covered woods by myself.
>
> Overall I have enjoyed the experience of doing this project. It taught me to just go with the flow a bit more than I usually do. It has taught me many new things and I cannot wait to start applying them to the real world.

Sam was one of a pair of identical twins. On first meeting, he and his brother seemed almost exactly the same physically, but in a short time, I learned to tell them apart. The school had placed them in different sections in my classes; this was the first major course in their four years here that they had not sat together in the same classroom. For the writing of the EA in particular, I think this was a good thing. It turned out that many of their writings about the same incidents showed quite different perceptions. From Sam's analysis:

> I believe I am becoming more confident in the things I do. I am more likely to say things that I believe and more likely to be myself. This is honestly the first time that I have recorded my feelings in such a way. I am not used to talking about emotions and things of that nature, but now since I have crossed this barrier I find it a lot easier to talk about these sensitive areas. . . . This project . . . identifies the

weaknesses that you have right now. For example, you can tell in the beginning of this project I was unable to open up, but towards the middle and the end I was able to open and actually record some of the weaknesses that were holding me down. I think talking about some of my fears and weaknesses made me realize that I am not the only one who has these drawbacks.

Sam was very connected to his family. His parents were both teachers, and they had instilled in him a respect for education. They both taught in the inner city and had spent conscious time getting their sons and daughter outdoors and educating them in how fortunate they all were to live as they did. Sam wrote pages and pages in his regression about memories concerning water. At age six he went fishing for the first and last time. While learning how to cast his rod, he accidentally hooked a seagull that was flying overhead and almost drowned it. With some help from nearby fishermen, he was able to release the gull. He wrote, "I felt very badly for the bird but I thought it was cool to see the bird up close." He wrote about ocean waves, canoeing, puddles, snow, ice. "I did not care if the water was blue and beautiful or black and bacteria-ridden. As long as it was water, I was happy." In his analysis, Sam wrote:

> Water can relax me, excite me, and make me happy all at the same time. When I am in the water it is as if I am in a whole other world. I do not think about any problems I have, I simply enjoy the rush that the water provides. I believe that the water will always be my place of escape.

While he didn't write about this in his Stage Four, he and I had several discussions about the protection of lakes and other bodies of water. He saw water as therapy and realized he himself needed it for his "preservation."

From David's analysis:

> Nature is a place I can go and be engulfed by observing my surroundings. When I think about these experiences [his regression], in every one I am by myself. I have had countless experiences with other people, but the ones that have the most special meaning are the ones where I am by myself. When I am by myself sitting atop a cliff, overlooking a river, the peace that comes over me is indescribable. Even though these experiences of solitude are becoming less frequent, I have learned more about myself than other experiences don't even come close to.

David was in the process of "surviving" his parents' breakup. The divorce had been messy, and as David was the only child still living at home, he had been pulled in both directions and had watched his mother go through a particularly difficult time. David himself had done some acting out, and as his senior year wound down, there did seem to be a lightening of his load. His sister had given birth to a baby. David took a week off from school to be with her and help with the baby, and he seemed to come back somewhat restored. He made some revisions in his synthesis section:

> Now when I think of the future I know where I want to be. The goals that I have set for me are not unreachable. My goals provide me with hope and the opportunity to succeed. Even though the future is unclear, I know that as long as I am enclosed by nice environmental surroundings things will always be okay.
> My dream is to have the money so that I can live on a ranch somewhere where the air and water is clean and I am isolated. This dream is reachable. There are places where I can live and be away from civilization. I would like to live like my sister. I want to be self-sufficient. [His sister and her husband have a small ranch in northern California where they live as self-sufficiently as they can.] By living this way I would feel that I am doing something that benefits the earth. I would like to do some sort of research so that other people can see what living responsibly is like and how easy it really is. My hope for the future is that the population of the Earth levels off. There are too many people on the Earth; however, it is a hard goal to accomplish. Hopefully the world will begin to understand the impact man has done to the Earth. By living in an isolated area, I can show other people that living this way can be done, just like my sister showed me the first time I ever went to the Wiki Up ranch.

When we had discussions about deep ecology at the end of the school year, David was the only student to see merit in the approach. The rest of the class thought the ideas "would never fly," that people would not be willing to give up so much of their consumer-oriented lives. David, in his quiet and unassuming way, kept reminding the class that drastic measures may be our best hope to save the planet. The rest of the class, though, reminded him of his, his sister's, and her husband's large trust funds that would make a life of self-sufficiency, modeled on deep ecology principles, much more possible for them than for the vast majority of Americans. David did not deny this.

David's regression had many examples of his digging—digging actually became one of his themes. When he became old enough to use a small shovel, he began his digging life, spending hours making holes in the lawn, planting and replanting. Even today, he loves gardening and especially loves the double-digging method, where the soil gets turned over twice by hand. He talked about how his parents did not discourage his digging and instead gave him areas to dig as he saw fit. His dad bought him cones and flashing lights to put around his holes. David decided to renovate the school garden for his second-semester ecology project, further connection with his analysis of his EA.

But I am going to relate a non-digging episode from David's EA because it shows how early some of the students' memories were and how from the beginning of the project some were able to analyze their feelings:

My earliest memory just happens to deal with the environment. When I was about one and a half years old, I was sleeping in my crib and a thunderstorm rolled through. I remember the entire experience. When it began to rain, I remember thinking how peaceful the sound of the raindrops were. To this day I love sleeping with the soothing rain in the background. Everything was going well until the storm began gaining power. When it started thundering and lightning, I remember being afraid because I did not know what was going on. The storm was incredibly powerful. I remember my mom taking me out of the crib and taking me into the basement. I was filled with so much curiosity that I started to cry because that came naturally. I was captured by the fear of not understanding the situation. The power of the storm added to my search for the answer. I wanted to know why the sky was flashing and why the clouds were making noise. My mom and dad didn't offer me a very good explanation. They said that G-d was bowling and the flashes of light was caused when G-d got a strike. My young mind believed this for a while until I found out what really caused lightning and thunder. I think that this is my first memory because of the great paradox I was in. I was afraid but at the same time amazed and soothed by the powerful rain. When the storm was over and I went back upstairs, I remember looking out the window and seeing how calm and peaceful the night sky was.

From Carl's synthesis:

All in all, this autobiography was one of the best projects I have ever done. It really made me realize how lucky I am to live on such a nice landscape and to go to a school where it's [sic] scenery is gorgeous. This autobiography really started to make me think in how many of the things that I have done, that were special to me, that involved nature in some way. I just really never realized it, until now. From now on when something happens to me that affects me in a big way, I will just think how did nature play a part in this. This autobiography really gave me a lot of time to reflect on my past. It was very exciting, looking back on all the things I have done. This project was by far the best project I have ever done and one that has altered the way I think towards the planet Earth.

Carl was one of the youngest members of the ecology classes. He was quiet and I worried about him finding his own voice in a class composed almost entirely of seniors. He never did become voluble in class, but he elicited a healthy respect from the rest of the class, mainly, I think, because of his good humor and willingness to do anything. He was a three-sport athlete at school and looked very healthy. He too, though, used the EA for catharsis and wrote about his lifelong battle with asthma. He actually went on a breathing machine every day and had been doing so since he was tiny. I had no idea that he was handicapped in this way. It is the outdoors that was one of his greatest problems, whether it was running, pollen counts, or cold temperatures. He wrote about how he had learned to deal with these situations and philosophizes: "All of the things that I have been through with nature were either positive or experiences I could learn from."

Many students wrote about making leaf piles. The following is an example from Carl's regression:

> The earliest memory I can recall of nature is when I was about six years old, outside of my old house in Mayfield Heights. It was a nice cool fall afternoon, and my older brother and I were outside raking the fallen leaves into an enormous pile. When the huge pile of leaves was complete, my brother and I flew into it. We must have played in the leaves for an hour and a half. We were jumping in them, sliding in them and tackling each other on nature's temporary cushion. This was probably the most exciting event that I had ever experienced up until this point in my short life.

From Steven's synthesis:

> My public life should be structured in a manner where I appear to be a very strong person mentally and physically with a great deal of confidence. However, underneath that bold individual I should have a sensitive side so I can put all things in perspective. Sensitivity is also a very important virtue. You can't channel all the negative aspects of your life out. You have to feel some pain in order to prosper. No pain, no gain. You have to have some emotion—you can't just be a brick wall and not feel the deep pain when someone for example dies. My physical body will always be in some kind of shape, I hope! Its connection with the physical environment will always be there. I love being outside whether it be playing a sport or walking outside. This love and passion for nature will last forever!

Steven was the other half of the identical twins. He was an athlete and came to this school to play football. He was taking this course because he needed the science requirement and chose this one because another member of the football team had suggested it to him. He had no interest per se in ecology. But by the end of the year, he had become one of the most outspoken members of the class and spent a great deal of time on all his ecology homework. He became very serious about ecological principles and often talked about them with students not in the class. That was great, because no one teased Steven or made fun of him behind his back. He was probably the biggest jock on the campus, and his involvement with the class did much to promote the course to the rest of the student body.

More from his synthesis:

> Throughout this assignment I believe I have changed dramatically. I really got in touch with my inner self and who I really am. The most important thing in life is not to forget where you came from. This project has opened my eyes to experiences I would have never remembered if I had not thought in such a deep manner.

... All emotional barriers were let down and I wrote what I was really feeling. . . I am truly thankful I have such a great family who has structured my life so well. For example, putting the right virtues in my head.

Steven's recollections of an elaborate treehouse his dad had made for him and his brother and the backyard skating rink his dad made every winter were told in such a way that it took me a while to realize that he and his brother did not live in the country. During one class when we were not discussing EAs, Steven just happened to mention that his treehouse had been demolished by the next people to move into that house. He said to this day when he goes by that house he feels pangs of loss for those days.

As often as I could, without getting tedious about it, when the students would share their regression experiences with the group, I would ask them about the environmental conditions of the place they were mentioning—how polluted was it, was it clean, was it safe—and whether those conditions contributed to their experience, positively or negatively. I was trying to begin a connection with preservation of place in their heads—if these experiences were so meaningful to you, is it going to be possible for your own children to have the same experiences? If not, what can you do about that? Do you *want* to do anything about that?

Another interesting thing in regression we were seeing over and over again was how many of the experiences involved nature. Even though that was part of the assignment, to limit experiences to those types, we were finding in discussion that those were indeed the most meaningful positive experiences. For some of the boys, seeing how significant a part the outdoor environment played in their early lives was eye-opening; they had not thought of themselves or those experiences in those terms.

From Clark's synthesis:

The process of creating this journal has been one of many new realizations about myself. Actually putting down my memories, fears, goals, needs, and expectations on paper has allowed a huge weight to lift from my shoulders. For many years I had missed my carefree innocent days of being a child without a care in the world. I knew that such a feeling was a great one, yet I had no clue as to how I could achieve such emotions. In writing some of my earliest memories, I have actually been led to feel as though I have traveled back to those warm sunny days when the whole world was a mystery that I did not take for granted. A part of my childhood has been rekindled even while I get older and grow up.

Data from the First Year's EAs

The world as "mystery not taken for granted"—Clark has captured the purpose of the EA in a few words. If somehow every participant could share that feeling in some measure!

Clark was a very gifted student. In the ecology group he was in, he was the catalyst, the point of abrasion that got everyone thinking harder and deeper than they would have without his insightful, incisive questioning. He had a good sense of how the world works and tried always to refocus attention on whether something would or could "work."

Clark carried this idea of child/mystery further into his synthesis:

> When I move on to the next point in my life and I think about where I have ended up, I want to feel like the curious child I was. I want to look around at my college life with surprise as I think about how much my life has become such a different thing than it once was. I want there to be mystery and curiosity as to where my paths may lead me. . . . If I don't feel right in a particular job, I am going to do something else until I strike gold and find my niche. I don't doubt that I will because this project has kept me from doubting myself. I have learned how to know and recognize my own thoughts and beliefs in a way I could never have put into words prior to ecology class this year. . . . All that I do in both public and private will be with passion and honesty. I want to live as a person who feels truly alive at each moment of life and who knows that every day is an adventure of people and places. In doing so, my physical body will be better connected with my environment as the emotional bond has set a foundation.

Clark was asthmatic and had to carry an inhaler with him. The inhaler had become such a part of his life that he didn't seem to give it much thought; if he had to stop what he was doing to use it, he did so with no fanfare or complaining. I think the project helped him realize how wonderfully he was handling this condition and how much control he had learned to exercise over it. There were many days when he felt unwell, but little stopped him from going outside, or participating in sports or classroom activities. His classmates always referred to him as the Spazz, because he would so quickly spazz out if someone said or did something he didn't agree with. He was almost always emotionally involved with what he was doing. There was a sense of energy field around him.

From Clark's progression:

> Assuming I will have the right to produce all the offspring my wife and I want, I want my children to grow up in a place that is not only safe in the way of crime and other ills of human nature, but safe from a tainted environment. I want to raise them in a place similar to Chagrin Falls—minus the spoiled, drugged out, wannabe nature lovers. Why shouldn't my children grow up in a place that offers the best of human life and nature. I hope that there will be places my wife and I will be able to offer my children that are not tainted with acid rain, pollution, pesticides, lack of forestry, and no safe body of water to be near. Maybe my hopes are idealistic, but if others of our generation want such a thing for their children, they

need to begin to think about what will make their earth a better place to live. Such a place I would like to go could be near a beach, perhaps one without syringes washing up to the shore.

From Clark's analysis:

> There is no definition of family that can fit in this day and age so I must say that I have sort of another family. A family gives support when any one of us is down and does not discriminate for ones [sic] thoughts, but maybe laughs at them from time to time. I remember when my parents were getting divorced and a psychologist told me I was depressed. He said that the reason I was always with my friends was because I was in need of compassion I wasn't getting at home, as if my parents were inadequate. The fact is, my friends have given me support that only they could provide. In just a few words, my friends have helped out when people have tried to fight me, been there for me when there have been family sickness, tell me when I'm out of line with a girl, get my back when I am being robbed, stand up for me when I have been wrongfully accused of hitting a guy's car that had run me over, allowed me to give them support during the bad times, gotten arrested with me for the first time in eighth grade, taught me how to kiss a girl properly, gotten me to stand up for what I believe, given me a shoulder to cry on, fought me to show how wrong I might be, given me an ice pack after fighting me, trusted me, and most of all just giving me great and happy memories.

Clark's parents divorced when he was in middle school, and he spent his early teens confused and angry. He had come a long way, and I think in some part the EA was a chance for him to give himself a pat on the back for coming through this period so well. He was a loving, affectionate, caring, empathic individual.

From Clark's regression:

> The earliest memory I have concerning an environment and the way it made me feel took place when I was about two or three years old. I was younger than four because my little sister had not been born yet. I was on a car trip with my parents. I believe we were heading to my grandparents' house, when we pulled over to a dingy little rest stop somewhere along the highway. My parents went to the bathroom, leaving me with my older sister Nikki, who was around ten years old at the time. We were supposed to stay in the car until my parents returned, but my sister saw something that sparked her interest. A group of people were standing on the sidewalk pointing at a grassy picnic area. They seemed very surprised, so of course my sister had to see what was happening. She couldn't leave me alone in the car, so naturally she took me along. We approached the grass and the closer we got, the more we began to notice a strange sort of coating over it. Finally reaching the picnic spot, we bent down to analyze a kind of shell that covered the entire grass patch. It appeared to be that of a large bug, but I didn't know that they were just shells. My sister did, however, and wasn't scared. She picked me up and carried me to the farthest end of the small field before it converged into a typical wooded area that can be found nearly anywhere in Ohio. My sister was so excited by what she saw, that she took off running around. Every step she took, she crunched many of the seemingly millions of shells. I on the other hand, was three

years old and was not impressed by what I saw. I believed that the shells were actually giant bugs that would sting me or bite me. The heat was beating down on my tiny little body and I began to cry. My parents quickly ran out of the dirty restrooms to see if it was in fact their own child bawling his eyes out. Their rescue, nonetheless, was cut drastically short as they joined the group of spectators in awe at the amount of shells. Eventually, my mom ran over to comfort me as I stood in my own frozen tracks. Shells were blowing onto my legs and shirt, making me cry even more. I was as scared as I ever had been. Thank the lord that my mother was soon holding me in her arms and putting me on the safe pavement away from the tiny monsters. My sister was still crunching around, and I was very scared for her. My parents explained to me that the shells were not real bugs but in fact only the outer layer of locusts. I didn't quite understand what they meant by a locust, but I felt better when we were in the car driving away.

Clark wrote often in his EA of new experiences, wanting to have new experiences, the excitement of doing something different, seeing something unique. When it came time for him to select an experiment for the second semester, he was not about to do anything suggested by me or his peers. He thought for several days before coming up with something unusual—a study of piranhas and carrying capacity, how piranhas live together, how they "learn" to hunt together, and so on. As with everything else he did, he became totally immersed in the project and spent hours studying two aquaria he and another student had set up. His project brought more students and faculty into the lab than anything else we did that year. There were non-ecology students who would even sneak quietly to the back of the room when I was teaching a class in order to check on the fish. Clark loved explaining his project to anyone who would listen.

From Arthur's synthesis:

This project made me remember some experiences I have had with nature, which at the time didn't seem like they would play a big role in my life. After thinking about these things in great detail, however, I realized that these things that I was seeing and experiencing weren't just vacation spots and sightseeing ventures. I realized that I am more a part of nature then [sic] I thought I was . . . this process made me realize that I wanted to live a life where I would have both nature and the city in my life. I could not fathom living a life without seeing a tree or being able to hear the sound of a bird chirping. Nor could I imagine living a life without gazing at majestic high-rises, gleaming like modern-day palaces, hearing the sounds of car horns beeping and hearing wheels rolling on the smooth pavement. Before this project I thought that I was the last person in the world to be connected to the earth, but now I realize I am just as much a part of nature's cycle as the hard working ant or the proud Arctic wolf.

Arthur had traveled to many interesting places: Thailand, where his older brother was involved with caring for refugees; the Amazon jungle; Hong Kong; London; and Australia and the Great Barrier Reef, to name a few. He had gleaned much from these travels and wrote extensively in his regression of these kinds of experiences along with the more mundane daily events. He had been at this school thirteen years and had not had an easy time of it. He had to work hard for his grades. His brothers also went here and graduated with significantly higher GPAs. Arthur had to live with that, too. He was friendly, cheerful, and a good conversationalist.

From Arthur's progression:

> I see myself living in the suburbs, fairly close to the city, with my wife and two kids. The house we live in is a red brick house, with two lights illuminating the walkway. I picture the house during the winter with icicles hanging off the roof and sitting by a nice cozy fire. I picture taking my kids on camping trips, and the whole family taking a vacation to places ranging from Florida to the Amazon jungle.
>
> Many years later my kids are through with college and are now living on their own. My wife and I retire to Hawaii, sitting side by side, drinking milk directly from the coconut, watching the sun set. There I spend the rest of my days living under the tropical sun, swimming in the clear blue ocean, and walking the clean beaches, with sand as white as snow.

✧ ✧ ✧

From Rhett's synthesis:

> High school can be classified as our glory days. We can't let them pass us by. We have to look at ourselves from the outside, like in this autobiography, and realize that life is the final performance, with no rehearsals. We can't take time on needless things. We have to put our fears and differences aside, and just live.
>
> By writing what you are about to read I was able to find an inner self that tells me that there is no time to be wasted. I have to take everything and be thankful for it, and I can't worry about the things that will be irrelevant in the future.

Rhett was a junior. He had had a shaky sophomore year and had garnered a reputation for himself as a modified troublemaker, doing things just enough on the outside of the rules to get in trouble but not so bad that punishment was ever severe. By the end of his junior year he had become a prefect in the school, a position of responsibility and much work, and had two major roles in school play productions. He was a serious member of the ecology class. There was nothing we did that he did not view with enthusiasm and concentration. Even though he was not a senior, the rest of the class depended on him for guidance in all matters ecological. In early December the weather had been unseasonably warm; green shoots were appearing in the woods and we could be outdoors without our coats. As the solstice approached, however, the temperature dropped precipitously. It was on one of those days that I made these field notes, field notes in which Rhett plays a major role:

> Friday, December 18, 1998, our last day of school before the two-week Christmas break, we have only 20-minute-long classes, and I decide to have my ecology classes make a Christmas tree for the birds outside our classroom. There are two white pines there we can decorate easily. I bring pine cones, suet, peanut butter, sunflower seeds, cranberries, dried apricots, thread, needles, wire, and tools to class and describe to the boys how we can make edible ornaments. Most everyone participates wholeheartedly—a few, like BJ and Walt, play the too-cool teenagers, but they end up sitting around the lab tables with us, sharing in the conversation. The day is beautiful. Snow is still on the ground from the storm of the previous night, where we had our first snow of the season. The tree begins to take shape after the second ecology class finishes, and I decide to have my anatomy class finish up the ornaments. (They, too, enjoy the process, and except for a few boys, everyone gets into the spirit of the work.) Rhett finished up his ornaments and as the class bell rings, he asks me if he can do a polar bear dip in the lake. He has come prepared—capilene, extra socks, warm pants—and wants to begin a yearly tradition of polar bear dipping before Christmas break. Tom, Gene, and I decide to go with him. I think Tom and Gene are not quite sure Rhett will go through with it, although Rhett has become known as "crazy Rhett" as these months progress. I go along mainly to be sure he doesn't die! But I too think it is a great idea and wish I had thought of it myself and that I was prepared to join him. We go down to the boat dock (no camera!) and Rhett removes his clothes down to his boxers, jumps off the dock, deciding he is definitely NOT going to get his hair wet, and immediately descends totally under the water. He comes up like a bullet out of a gun, a look of awe on his face, and we, who are holding some of his clothes, quickly begin to dry him off. After a few seconds, he can speak, and his cheeks are wonderfully flushed and he is excited. We all decide that next year, Tom, Gene and other graduating seniors who will be home on college break, will come up to school this last day and we will all jump in together. Rhett declares he will make t-shirts to commemorate the occasion. I not only want this to happen for the continued synergy of the group but also to reinforce in these boys who will then be off in a different world away from here a re-bonding with the environment—especially if they have been away from it for awhile with all the new stresses on college freshmen.[4]

Rhett had spent almost every summer of his life on Nantucket. Since becoming a teenager, his parents had allowed him to spend the entire summer on Nantucket with his brothers—no parental supervision. He spoke glowingly of the independence of those summers and how he and his brothers had become part of the Nantucket "regulars." He worked the same

---

[4] We did indeed do a second annual Polar Bear dip the following December, and Rhett and I jumped into the water together. Approximately 25 other students joined us. Tom and Gene were not yet home from college, so they were unable to be with us.

job there every summer and had become friendly with the local islanders. He related many ocean stories in his regression, but he also used the EA as a chance to explore his fears, of which he had many:

> It was during the Olympics when Greg Luganis [sic] hit his head on the diving board. From that moment on I am still afraid to do any sort of flip off any diving board. The same is true with trampolines. This kid I knew sprained his neck one night doing a flip on a trampoline. Now every time I jump, or see someone attempt to jump, I cringe at the thought of them hurting their neck.

Or,

> About a year ago my mom's friend's face got stepped on after she fell off her horse. Last month I was riding and that thought came into my mind. In my opinion horses have a sixth sense and can detect fear. The horse that I was riding was one that I was not accustomed to. It sensed my fear at the moment I thought of my mom's friend. All of a sudden it started bucking.

Or,

> Hollywood shapes who we are no matter what anyone says, for instance, the movie *Jaws*... Until this summer I could not go into the water alone... I am still afraid when I walk in the water that I will be eaten by a great white... Later in my life, when I was in third grade I watched Freddy Kruger. From then on I was petrified to sleep without the light in my closet on. I would go to bed only after I would check behind all of my doors, look under both of my beds, check the bathroom, the closet and any other place where a serial killer could hide. I even slept with this Swiss army knife in a drawer nearby just in case I had to be on the defensive. When my parents would leave me alone when I was young, I would go into their room, lock all of the doors and have the phone and the alarm trigger button by my side. My parents would come home and the alarm would go off, not even waking me up. They would then have to pound on the door before I unlocked it, being sure that it was them, not a mass murderer, behind the door.

Or,

> One time, when I was biking, I flipped over my front tire. I am always afraid I am going to do it again. On the topic of bikes, my brother was riding and for some reason his front tire fell off and he got scraped up pretty badly. Now I always check my front tire to make sure that it is really tight before I start biking.

Or,

> Once I read this article that you can choke on your own throw-up and that said people die as a result of suffocation. Whenever I am sick, I force myself to sleep on my stomach so if I do throw up I will not choke.

Or,

> When I water ski anywhere, say in Nantucket, I hate when the boat doesn't come back at once [when I have fallen down]. I sit in the water with my hands and feet out of the water so that if a shark or Muski [muskellunge] would swim by they would not take off my fingers, toes, arms, or legs. I literally shake in the water because I always think that I happened to fall in the part of the water where the sharks feed and where they would be lurking to find their next meal.

Or, my personal favorite:

> My mom came from a family with four brothers and four sisters. She was full of stories that have changed my life. One of the many is about her youngest sister, Amy. While Amy was very young she was outside drinking soda from a can when all of a sudden a bee flew into her can. She had no idea it was in there so when she drank, she swallowed the bee and it stung her in the throat. Her throat swelled up and she was rushed to the emergency room. She turned out all right but could have died. After my mom told me that story I always drink my pop differently. I put it in my mouth, then filter it through my teeth to make sure there are no bees sneaking into the can.

I cite these different regression sections about fears because if you met Rhett, you would not believe any of these stories. He exuded confidence, especially when he was outside. He was the first to test the lake for safe ice. He erected a rope that traversed the lake and hung a pulley from the rope. That way he could connect himself to the pulley by way of a climbing harness and maneuver over the lake so he could hover over one of the beaver dams and take pictures. He jumped into the lake, no matter what the weather, quickly stripping down to his underwear, jumping in, breathlessly rising up, throwing his clothes back on over his wet body and declaring, "Who needs drugs? That's the best high in the world!"

Another regression story:

> The first trip that I remember is a small one but has had an impact on my life forever. It was the yearly trip to my mother's uncle Morgan. He was two hours away from where we lived in Winnetka, Illinois. Although I do not recall the actual happenings when we got there, I do remember the car ride there. These two hours has been the basis of my time scale for the rest of my life [sic]. Even now, I am sitting in the car, speeding over the highways of Indiana, and I know that there are two trips to Uncle Morgan's left to go until I get home.

Rhett decided to do an in-depth study of the two beavers that live on our lake for his second-semester ecology project. Besides using the rope he hung to take pictures, he spent several nights camping out on the lake to observe the beavers at their most active. (Beavers are nocturnal animals, but for some reason, our two beavers did sometimes move around during the afternoons. They were very skittish, though, as one can imagine when surrounded by a campus of four hundred teenage males. While we would catch quick glimpses of them gnawing at trees or swimming during the daylight hours, there was much more activity to record in the dark.) I accompanied Rhett on two of these sojourns, the first time because I thought he might be more afraid than he was letting on, and the second time because I knew he was much less afraid when I was there. We did witness some amazing behavior and experimented with using a tape recording of falling water to coax them out of the den.

As an outgrowth of the beaver project and the EA, Rhett became an avid helper to the fellow on our campus who is in charge of the trout hatchery and maple syrup production. This commitment for Rhett would have been considered impossible just a year before. Today, many of the ecology students view Rhett as the most ecological of the group.

From Walt's synthesis:

With the difficulty I have had voicing my deep thoughts, this Environmental Autobiography has given me the chance to finally tell others of how I feel, without the awkwardness of voice explanation with another. Though I still haven't even begun to express my inner thoughts, I have at least cracked open the door by putting a few ideas on paper. In doing so, I have also discovered that I am a decent writer, contrary to what I used to believe. This Environmental Autobiography has seriously affected me, making me think about possibly doing this on my own. Before I never saw writing playing a role in my future, but now, with the impression that this [project] has had on me, it would be a shame if it wasn't.

From Walt's analysis:

I am often plagued by deep thought, preventing me from falling asleep or paying attention. By asking myself questions which I can never find answers to, I provide myself with an endless amount of finding out. One question I think about is are we better off with the evolution of science and technology? Of course we have learned fascinating things, especially in space, and we can travel to all points of the world in a matter of hours. But we also have lost the opportunity to be completely free, not freedom of speech or religion, but to go and do what we feel on a whim. We are now troubled by an enormous amount of responsibility. As intriguing as discovering the actual information of the creation of the Universe is, are we better off having faith in something that we really know nothing about, even if we are wrong? And as nice as it is having a guide and a beaten path all over the world, might it be more fulfilling to create your own path? It is almost as if science and technology have acted as the spoiler to the Earth's surprises; we know what is around every turn. I believe this is a key reason why many people have lost touch with engaging in nature because they may feel there is nothing left to be explored. To our civilization the Earth is like a parent, when we were young we were in awe of its vast lands, and amazed by its power. Now, we have grown up, the Earth isn't so big anymore, in fact we are taller and have harnessed its power. By building skyscrapers and flying in airplanes we can look down and feel like we are in a position of control. And now the withered and aging Earth is tossed aside, like an old toy.

So many students speak of this freedom that exploring the wilderness can bring. At this stage of their lives, they are so close to "breaking out," leaving behind the rules and confinements of living at home for what they imagine the college life to be—a chance to make their own rules. Walt's theme throughout his EA was one of exploration, whether it is a small exploration in the local park or hiking the Pacific Northwest. In his progression, he wrote of future explorations and detailed situations where he comes across American consumerism in the remotest of areas. In one such story, he mentions meeting a man at a pub:

> He begins spewing information about how alcohol was a staple in the locals' lives and how difficult it is to work your way out of the village. This is when I realize, once again, how diverse our world really is. I often fantasize about living the life of a farmhand and enjoying the simple pleasures of life in a small town, as I have seen what the urban world can offer. It is interesting that others without my opportunities wish to be in my shoes, while I wish to be in theirs.

Walt at eighteen was a rather sedentary fellow, content with discussing what was going on ecologically rather than getting out and checking for himself. He did love to hike and wanted to do his second semester project on how hiking trails are formed and marked and how decisions are made to produce new ones and abandon existing ones. I thought this would be an excellent chance for him to see the politics involved in trailmaking and arranged for him to meet with a member of the local metroparks whose main job was working with trails. This fellow proved to be very busy, and Walt had to rearrange his schedule several times in order to get together with this person. Unfortunately, Walt lost patience with that, and the project never really got off the ground. He instead ended up repairing a section of one of the trails on our campus. It is a good indication of where Walt was at that time intellectually that he blamed the metroparks person for the failure of his initial project. He planned on studying Environmental Science at the University of Michigan the following year, and I hoped he would soon realize that the solution to his problems was not necessarily around the first corner he took. He had a sincere interest in figuring out how to save the planet, an interest that I don't think was at all apparent to him until he started writing his EA.

From Walt's regression:

> When I was five years old, I was restless. One day while my parents weren't near, I can't remember why, I decided to climb the tree in our backyard. The tree was

pretty large, all I remember was that it was thick and gray-colored. It towered above our full-size house. It was warm outside, but I had a long-sleeve T-shirt on, and pants. As I began to climb the tree I didn't think of any possible consequences. Since I was invincible I went from branch to branch with ease. Birds were chirping. I now was about half-way up the tree, and I looked down at the lawn below. I proceeded up the tree and I then began to wonder how I would get down. But that didn't matter at the time, I was climbing the tree. When I reached the high point I looked around not feeling any bit frightened. I was perfectly calm. It was a gorgeous day. I looked out of the sphere of branches and leaves that the tree created, and saw my backyard neighbor's house. The windows were open. In my line of sight there was so much shade and so much light. On the other side I could see down in through the large window of my porch, even though I was above it. Then I looked down. I saw the blue, red, and white jungle gym and wondered if I could leap down on top of it. The leaves rustled in the wind. Alas, I realized that probably wouldn't work very well. I found myself in a sitting position and relaxed for a little while, couldn't stop looking into the yards behind the fence of my backyard. One with a plethora of pine trees, the other, a small backyard with a patch of lawn and a house that seemed to spring from it. I found myself up in the tree for a while, so peaceful. It had to be late-morning time, workers laboring in yards everywhere. Time to come down, don't know why, it's just time. Looking down the whole time I just proceeded down with ease, though teetering on disaster with every move. I jumped down, and decided that I didn't want to go inside, so I laid out on the grass next to the tree, so soft.

This is a story where the reader expects the broken arm and the irate parent remonstrating to him to never climb that tree again! But no, the story is one of exploration and discovery that has stayed with him for many years and will now stay with him for many more because he has written it down and given it form. He writes of several events in a similar vein, each time experiencing a sense of peace and calm that he cannot forget.

From Emery's synthesis:

This process has been an invigorating experience for me. It has served as a way to unlock certain intellectual and emotional gates in my mind. I am a totally different person now because I have a new understanding of the outside world around me. When I first began this project, I had mixed feelings about the whole thing. The idea sounded great but I wasn't sure how successful it would turn out. My doubts were further enhanced when, in the first stage of the project, I had great difficulty remembering the events of my past. I can remember the feeling of frustration I had when I struggled at first to put some of my feelings and experiences into words. I wasn't sure if I was even on the right track.

This was Emery in a nutshell. Emery was an intense, very smart, very private individual. He entered ecology unenthusiastically. He needed credits for graduation and this class seemed to best fit his schedule. When I first began teaching the class about critical literacy, he was one of the first to grumble. Well, how many points will the essay be worth? Are we going to have a quiz each day on the reading? Do I need to memorize all these dates? And then when I introduced the EA, he was upset. He couldn't remember anything, he said. He would flunk the project. He didn't like analyzing things. Wasn't there something else he could do instead? I asked him to try to approach the project with a more positive attitude and just start writing—write whatever would come into his head. He showed me his first draft in total isolation—he didn't want anyone else around when I would read it; he was sure it was not at all what he was supposed to be doing. But in that first writing there were glimmers of stories, and we talked about them together. I asked him to go home that night and write up the stories he and I had discussed that day after class. He did—ever the dutiful student!—and added one more. He brought them to me the next day, and I told him, bingo! That's it. Try some more. And the writing just kept coming:

> As I began to spend more time each day thinking about my past, the memories started to flow out and everything became easier. I think the problem was that I tried to sit down and start writing immediately, figuring that things would just come to mind. But that was not the case as I soon found out. It took a little bit of time where I just sat and thought about my past. After I did that, it became a joy to sit and record events in my life. It opened me up to the environment and helped me form the foundation for what I come to expect in the future. . . . I have learned to appreciate what I once took for granted. I have come to recognize that the environment is not just a place where I live, but rather it is a main factor in the shaping of myself and the events in my life.

From Emery's analysis:

> It seems as though it is in my nature to worry constantly about things. Everything I do has to be done to perfection . . . everything in my life has to be organized (not necessarily in a neat fashion although most of the time that is the case) and on schedule. For instance, when I gave the speech for the introductions of the hockey team, I worried about some of the smallest details. Some of these details include the lighting on stage, the height and width of the podium, my posture, and whether or not my buttons on my sportcoat would clatter against the podium. All of these things had to be perfect for me and thus it took hours of rehearsing to get it right. I need to teach myself how to harness this intensity.

He did relax more as the year went by, and as the EA began to wind down, I found I could tease him and he would tease me back. He was more willing to take criticism from the rest of the group. He learned to write beautiful observations in his lab journal. He became the biggest proponent of our meditation sessions, actually "accusing" me once of causing him to play poorly in a hockey game because we had meditated earlier that day in class and he was much too loose to play well. By the end of the year he was confiding in me about his dilemmas in college selection and his problems with his girlfriend. And he was now talking about his writing. Prior to the EA, he felt he was a horrible writer. When he came to the school as a freshman, he had considered writing as merely a way of getting homework done. When he encountered the strong emphasis our school places on writing and the criticism he was getting for his writing, he felt he could never figure out how to be a good writer. In one of the drafts for the EA, I wrote him a comment about how pleased I was with the craft of his writing. He came to see me that day and asked me if I really meant that. I also heard from several other teachers how much Emery's writing improved his senior year, and I attribute that to the EA. The EA gave him a vehicle to get his thoughts on paper, and his thoughts were more literate than he at first realized.

Because of Emery's initial difficulties with regression, I was worried that he would really struggle with progression. On the contrary—I think the picturing session we did with meditation helped him particularly, and he wrote describing his pictures. He wasn't only fantasizing, though:

> I often think to myself about the possibility that things won't work out the way I want them to. Although I am very anxious to see what my future beholds, this feeling of doubt often makes me want to stay where I am, and not move on. This feeling is very hard for me to put into words. The closest anecdote that I can think of is the way I feel right now about going to college. People often ask me if I've had enough of high school, and am I ready to move on to college. But I can never give a definite answer because I have mixed feelings about the situation. One part of me wants to go on to college and explore new limits. But because I am not yet set with any colleges and I am not sure where I even want to go, the other part of me wants to stay in high school for the moment. In some ways I wish I had another year to decide where I want to go and what I want to do.

Emery was eventually able to remember back to when he was two years old, visiting his aunt in California. He wrote in his regression:

> I do remember that she had an enormous house set high upon a hill. I remember noticing the grass, which on top of the hill was short, well-kept, and bright green. But as the hill sloped downward, overlooking a narrow road below, the grass was much longer, and of uneven length. Most noticeable, though, was the difference in color. This grass was more of a brownish-green color. The sun seemed to have dried it up, making it look stiff and uncomfortable. I remember feeling scared that if I got too close to the edge of the hill, that the grass would reach out and engulf me and I would be lost forever in its tallness. But then there was the safe, confined feeling that I got from the grass at the top of the hill. It seemed much more comfortable and I was at ease with myself.

He used this tack with many of his regression stories, describing a dialectic pull that he would need to resolve.

From Carson's synthesis:

> I began this paper by explaining how hard it is, for me at least, to compose and share so many of my memories. I wanted to end with a quote by author Paul Watkins, which I think supports this hardship. In the last lines of his autobiographical book *Stand Before Your God*, Watkins writes "So we may never return [to Eton (his high school)], but far into the future we will still look back, until we understand the questions and have put them into words. Then we will stay silent, because this knowledge was not meant to be shared. We will keep it hidden like a pearl in the oyster of our gray and aging hands."
>
> I heard someone ask Paul Watkins a question regarding what he meant in this passage. Watkins answered that what he was trying to convey was that what was important in life was not the ends, but rather the means. What makes each individual an individual is the story behind how they became who they are. It is this story that we keep silent about because in the end that story is our own, and every body has a different one they must discover for themselves.

Carson was one of the most introspective students in the group. He was an ecologist long before he entered the class. He grew up with a mother who was active in local environmental programs. He spent a great deal of his time outdoors. He was a wonderful writer; his observations notebook was a joy to read, and his lab journal on experiments read like a well-written novel. He planned to pursue the Arctic studies program at Bowdoin College in the fall.

From Carson's analysis:

> I really find it scary that with all my education and knowledge regarding the environment and its destruction, both learned and experienced, I find myself doing little about it. I know about our pollution problem, yet I am still content and would rather drive a car to my destination. I know and have experienced our deforestation problem, yet I don't do anything directly to stop it. I feel that I do lead a relatively "environmental" life, however it is a scary thought that with all I know about our problem, I still contribute to it in many ways.

Carson had spent all of his teenage summers canoeing in northern Ontario. These were intense experiences for Carson, one of the reasons he voiced concern in his synthesis about keeping silent about these stories. He wrote his main college essay on these canoe trips and reprinted some of it in his regression:

> Keewaydin is a camp unlike any other I know. It is a place that puts you together with nine other boys your age and two extraordinary staffmen, then it lets you loose in the Canadian wilderness for six weeks. During this time you are tested in many ways. Every day you are pushed to the limit physically on long arduous lake paddles or overgrown and bushwhacked portages. You must keep yourself healthy mentally as well while you come to terms with a completely different world and culture. Over my five summers, I have built a solid base of skills which are necessary to navigate high choppy waves on windy lakes, or technical white water. Through the duration of the trip, you must keep alert and prepared to insure the success of the section. It is under these demanding conditions that the best and most interesting relationships are formed. . . . One of the major challenges of a Keewaydin experience is to learn about the personality of each sectionmate, and through your everyday interactions emphasize his strong points, and fill in for his weak points. I, for example, am a skilled canoeist with great energy and a large measure of good cheer. I was quite often the one who was asked to retrieve an overturned boat, as well as raise the spirits on a long, cold, and rainy day. I was not, however, asked to cook much.

The last couple of summers, Carson had more involvement with the Cree Indians they would encounter on their voyage. The Cree would invite the canoeists to their homes, show them how to make crafts, and spend time on the river with them. On several occasions I asked Carson to write more about those experiences, the encountering of a dying culture and the knowledge he was gaining about how that culture operates. The first few times he politely told me he would try, but days would pass and there would be no further paragraphs on the Cree. I kept nudging and finally he told me he couldn't—no one would understand, he said. "I can't do it, I don't know how to do it. Somehow writing about the experience will change the experience, and I don't want to do that."

From Carson's introduction to his regression:

> It is always hard to try and share your close memories with others. It is hard first of all to remember the accurate details. It is even harder to try and put those details into words which are worthy of your experience. The hardest part, however, is trying to convey to others the meaning which these events have to you. I often find it frustrating to try and explain to someone a story from, for instance, one of my canoe trips, when that person was not there to experience what I did. To them my story holds little significance or meaning, which is the frustrating part. It is these stories which make me who I am.

Sometime during the first quarter of the ecology class, Carson confided in me that he would love to build a wooden canoe. There is a fellow at our school who has built wooden strip canoes, and he had the templates for the canoes on our campus. Carson wanted to know if it was possible for him to build one on campus using these templates. I was determined to make that happen. During the winter months he and I explored getting the right kind of wood (red cedar, today only available for purchase in western Canada), the epoxy, fiberglass, and other materials. He commandeered two friends to work on the project with him, and beginning in early April, the boys, with the adult canoe-builder giving them advice every step of the way, cut their strips, steamed them to get them to curve properly, and began the laborious process of gluing and nailing each strip in place. By the time May came around, I was arranging in-school time for them to continue working, and the boat was really taking shape. By the end of the school year the boat was finished and would be taken for its first real trial run up into Canada that August for a week-long trip. I am not sure this project would have even come up if it weren't for the EA and Carson's struggle to write his regression. The canoe project became a project for most of the senior class—boys would stop out to the shed where Carson and his two friends worked during their free periods and cheer the boys on; faculty were visiting to see for themselves how beautiful the boat really was. Carson will have this boat for the rest of his life (his family paid for all the materials), a constant reminder of what he was able to do the spring of his senior year in high school.

From Larry's synthesis:

> There are so many different things in my mind that I don't know where to start and where to stand. Many people don't realize it yet but we are killing our universe. We put so many toxins in to our environment that it is just killing everything along with ourselves. Sure, you can hide the damage you do by doing something good, but it still doesn't take the fact of the matter away. We as the future have to take a stand and save ourselves from the deaths to come.

Larry was one of the youngest students in the class, but he had a brother who was a senior and to whom he was very close. This gave him a major advantage with the rest of the class, and from the very beginning he was just part of the group. His thinking, though, was definitely not at the same level as the seniors, which will be apparent in his writing. He was a very ebullient fellow and was particularly enthusiastic about ecology. Both his parents had environmentalist leanings, and Larry had learned much from them and the way they lived their lives. Both his parents were also wonderful artists, his father a well-known sculptor in the local area, and Larry brought that artist's eye to our observations. He became the class photographer and took great pictures of class field trips and different projects the students were engaged in second semester. He himself became very involved with building a waterfall on campus at one of the side streams on the property and spent countless hours designing it and building it. He thought it would be easy to divert the stream to do what he wanted, but he quickly learned that streams have minds of their own, driven by the principles of physics. He figured things out, though, and created an interesting, artistic fall.

From Larry's analysis:

> My goals are high, my standards are too but these don't keep me away from trying. "Where there's a will, there's a way." Seven of the most meaningful words to me. You can never shoot your goals to high [sic]. You can do anything and be anything if you are truly dedicated. I intend to be somebody, a father, a husband and a good friend. I don't think these goals are too out of reach. If I work hard and really dedicate my life to something I love, I will be that somebody that I am shooting for.

From Larry's progression:

> My mother says that I would be great in a talking job. I was blessed with skills and can persuade someone to do almost anything. I think that the ability to talk is a great gift because talking is a great way to communicate and if you are good at it, you can make a living with it.

From Larry's regression:

> My father, coming from Germany, was a woodsculptor and a part time grounds keeper at [a local private school]. Coming from a family of the great outdoors he taught me a lot of things about nature and my surroundings. I remember sitting with him in the school's enclosed tractor plowing snow with him at around five in the morning. When we were done my brother . . . and I would make huge snow forts out of the snow mounds, digging beautifully shaped tunnels from one end of the fort to the other with get away slots in-between each end. My father would also build a snow fort opposite of ours and we would have snowball fights until someone got hurt or my father was too cold to go on. I have always loved the changes of season and the changes of weather.

Larry mainly used his EA as a chronological journal of his childhood but found it difficult to go much beyond the reporting of these stories. He had been severely beaten his freshman year by a group of boys he passed on a downtown street. I could not get him to write about that experience. It had been very traumatic for him and he just wanted to forget it.

From Tom's synthesis:

> When this project first began, I thought that there was no way that I could write a twenty page paper, especially one about myself. Honestly, I hated the idea of writing an autobiography. I thought Mrs. Doerr was as crazy as [Rhett] for jump-

Data from the First Year's EAs

ing in the lake on that snowy day, when she assigned it to us. Nevertheless, I quickly learned that Mrs. Doerr wasn't making us write this as a punishment, rather she wanted us to touch on a part of our lives we rarely associate ourselves with: how we are connected with our physical environment. Now let me tell you a little something, I had a great time writing this paper as I remembered my storied past. I graciously thank her for giving us the opportunity to embark on a project such as this. Mrs. Doerr I am not trying to earn extra points here, when I say from my heart and most definitely from the whole class—thank you. I have never had so many fun times in nature as I have had this year in ecology. The observations and the walks in the woods are tremendous experiences and a great hands on- or being a part of- approach to learning. In summary, this year's ecology class with all the characters . . . will always have a place in my heart. You guys make me part of who I am, well you make me COMPLETE! [Tom then adds a quote from John Lennon] "Having fun is a part of life, you know, and without it, there's nothing but fear and insecurity."

Tom ended up writing much more than a twenty-page paper for his EA, despite breaking his arm shortly after we began the regression stage. He was in a full arm cast until the synthesis stage, but managed to type one-handed (his nonwriting hand), get his mother to help him type, and on occasion I would type what he would scratch out on paper the night before. Once he was out of his cast and we were winding down into the synthesis stage, he spent days rewriting and adding new material. He was the one person throughout the project who wanted to share everything he wrote. He was always so proud of his work and wanted critique from the rest of the class on his sentence structure, word choice, and so on. When he was finished writing, he added photographs, quotes, and the lyrics to "Imagine" to his project. He then paid to have it specially bound.

From Tom's analysis:

So you've had enough of my stories and you want to know who the real [Tom] is. Well, I see him as a hard working, easy-going fellow, who at times is a little goofy. For the most part, though, he is happy except when he gets a little stressed out from the daily school work, which tends to make him forget things sometimes. He really doesn't like school that much, but he has learned to accept it. He wants people to be happy and if they are not, he feels his job is not complete. Overall, he is extremely pleased to be alive and enjoy the richness that everyday life and nature has to offer. Each new experience, good or bad, has a place inside him where he learns and grows to become, hopefully, a caring, passionate, sensitive man.

From Tom's regression:

[talking about the past summer's vacation in Hilton Head] . . . getting up one night about three in the morning, grabbing a lawn chair and going to the beach by myself. I sat there relaxed, a Cuban in one hand and a Pepsi in the other. It was great, I was all by myself, alone, yet not alone, it seemed as if the millions of stars, the harvest moon—which radiated the beach—and the palm trees I saw that night all belonged to me. "I am Spartacus," I yelled. It was a great moment in my life: I realized that man can coexist with nature and live to cherish its beauty. That is of course, if we are responsible and not arrogant towards our environment. Thanks, Rachel.

The Rachel Tom is referring to here is Rachel Carson. During the writing of the EA, we had also been reading *Silent Spring* together. I assigned quite a few critical essays on different parts of the book during those months, and the students came to be very familiar with the author and her life. They all came to call her Rachel. Tom, in particular, was able to contextualize Rachel's life with the fifties (he was taking a course on the fifties and sixties at the same time) and with her gender, and he often talked of her bravery. I think part of that goes toward explaining some of what he wrote in his progression:

> I dream of breaking conformity—living by my own standards and being radical. . .
> I feel the need to sacrifice myself, lay my body on the line for a worthy cause. As days pass, I have begun to realize more who I am and look up to the people who are helping me find myself
> What I mean by being radical is being somebody who does something that they believe is morally just, even if it is something everyone else can't understand. Maybe it involves a problem in society, maybe it doesn't. I can't pin down an issue right now that I can picture myself in a role such as this. Luckily, there are no wars going on at this time and everyone has the same rights. However, I would definitely like to help people by joining the Peace Corps, or partaking in something of that nature.

He writes in his synthesis portion that he will put his EA into the bookbag he will take to college next year and reread it when he is missing his family or stressed about school. I hope he does—I particularly hope he rereads those sections about his choosing the "radical" life, rethinks his statement about everyone having the same rights, and begins to find an issue with which he can identify. He would be excellent in that role—he has leadership ability and the energy and enthusiasm to carry something out.

His regression section is filled with stories about his close family ties and his many friends. He also writes freely of his fears and how he learned to deal with them, at least some of them. He is one of many students who wrote about the Baby Jessica incident:

> I am not sure how old I was when Baby Jessica fell in the well, but I'm pretty sure this scarred me for life. Whenever I went somewhere, like Turtle Park, or somewhere with my parents, I made darn sure I wouldn't fall in a well or the ground wouldn't open up on my feet. I even remember praying for her with my mother that she would live. This incident was about the time my sister, Jessica, was born. Her birth had made me a little protective, as most big brothers should be for their sisters. I remember seeing pictures of Baby Jessica on TV and in the papers and

thinking I can't let that happen to my own Baby Jessica. When they finally got her out and everything was fine, I recall a poem my dad wrote that went something like this—Ding Dong Bell, Jess is in the well, who put her in? Mean Vikram Singh, (a friend of mine) who pulled her out? Brave [Tommy] stout.

From BJ's synthesis:

This assignment at first was not well received by my class or by me. She [Mrs. Doerr] was asking us to write a paper over thirty pages long. Her way of thinking was ludicrous to me at this point. I put the folder in my locker and forgot about it until two weeks later when we were reminded that part one of four parts was due the coming Monday.

That Sunday night, I reread the assignment and got to work. I sat at home and started typing. The first few pages were surprisingly fun to write and the more I started typing the more I remembered about my life. I found myself writing with ease and enjoying typing.

This was only the beginning. This assignment was worked on by me religiously for the following three months.

Indeed it was. By the time BJ was finished he had 72 pages in his regression alone, over 125 pages total. Early in his synthesis, BJ wrote about the difficulties he had had in school the previous year and how he was attempting to begin this year (his junior year) on a better foot. He had not been particularly happy with me or the class prior to this assignment, thinking that somehow I was out to thwart his job as starting quarterback for the football team by planning large assignments and tests right around his most important games. (Throughout the year, this feeling would crop up, that I had something against him and would deliberately make his life difficult.) What drew BJ into this writing was the EA itself:

I have learned what makes me the way I am today. The first and most important thing that I have achieved via this project is the ability to think in a clearer way. . . . I never really was one to dwell on the past, but on occasion I would try to figure out why something's happened the way they had [sic]. Never have things been so clear about what has happened to me, or what I want to do with my life. Completing this assignment, I feel as if I let a tremendous weight off my shoulders. Not because of completing the lengthy assignment but because I feel better about the way I live.

He makes a wonderful connection with work he had been doing in English class:

We were reading . . . Walden. . . . In this book it talks about his ritual taken from an Indian tribe to "busk." This is when he takes all that he doesn't need and throws it into a fire. He does this in order to make his life and surroundings more simple.

> This assignment has put me through the "busk." As I said earlier, I think clearer, thus showing a more simplistic view of my life. With this simplicity I feel a strong burst of motivation in all different areas of my life.

This, too, was no exaggeration on BJ's part. He went from very poor grades his sophomore year in all his subjects to solid scores throughout his junior year. He didn't lose that motivation at the end of the school year when so many students find it difficult to study for exams:

> For me I must try to stay on top of everything and know who I am around at all times. If I know what type of situations I can trap myself in, I will be able to figure out how to respond appropriately.

BJ's regression became so long because he took each story he wanted to include and made it into a story complete with dialogue and detailed descriptions of setting. He imagined how the conversations might have gone and described others' emotions as well as his own. Many of his stories are very funny; they tell of his naïveté in some natural setting and how he responded because he didn't understand yet how the surrounding environment worked. For example, the time when he was five years old and playing a game with a friend at her home, and he learned about the strength of river currents. They had used the girl's mother's jacket as something to hide, and when BJ placed the jacket too close to the river, the river swept the jacket away and carried it quickly downstream. The jacket was recovered several days later by a neighbor, and it was only at that point the owner of the jacket realized the coat was even missing! BJ had said nothing, and his friend was distraught, thinking she had somehow misplaced her mother's coat. BJ not only learned about river currents; he learned how much trouble he could get into with what he thought at the time was a simple untruth.

At some points, BJ used his EA for catharsis too. In fact, in his analysis he categorized those parts under three headings: drugs, race issues, and unfair stereotypes (especially regarding athletes). He worked through these topics with his writing, and one could see his ambivalence and then the strong values orientation he had received from his parents.

During the second semester, when the writing for this project was ostensibly over, BJ continued to keep a file of new stories he remembered for his EA. At the end of the school year, students can apply for writing grants the school offers for the following year to students who show a particularly strong talent and desire to write some extended piece of work. BJ applied and received one of these grants for his senior year. Without his experience with the EA, this would never have happened. Up to that point, BJ had never considered himself a writer or even a person who tolerated writing. Now here he was soliciting writing projects on his own, excited about the coming year not only athletically and socially but academically.

Summary

> The question is not whether we will live in artificial places but . . . whether the places we live in encourage a sense of belonging.
> —Chet Raymo, *The Path: A One-Mile Walk through the Universe,* 2003

I chose these excerpts from the nineteen students from the first year (one student did not complete his EA due to severe family problems) in an attempt to show how the EA had affected their lives. I will look later at how the EA drove other parts of that year's ecology curriculum. Throughout the hundreds and hundreds of pages I collected from these students, I could not find one negative section to relate in this chapter. I wanted to show the idiosyncratic effect the EA had on my students and yet demonstrate the collective goodwill they felt throughout the work.

By the end of that school year, I had become so enamored with the success of the EA that I began to get nervous about next year's class. Would they take to it in much the same way? How would my previous experience with it affect the way I presented it to the class? Would I be able to kick off the project with the same sense of experiment that I had the first time, or would my enthusiasm get in the way of the class's finding its own rhythm with the work? That is the topic of the next chapter.

Chapter V
Narratives from Subsequent Years' EAs: "Not Till We Are Completely Lost or Turned Around Do We Appreciate the Vastness and Strangeness of Nature"

> Pedagogy is the transformation of consciousness that takes place in the intersection of three agencies—the teacher, the learner, and the knowledge they produce together.
>
> —David Lusted, "Why Pedagogy," *Screen*, 1986

The narratives presented in this chapter represent sections of EAs from the second, third, fourth, and fifth year's classes (approximately 115 students). Before I relate excerpts from the students' writings, I will begin with my overall impressions of these subsequent years, the similarities to and differences from the first year.

Impressions of the Second Year's Ecology Classes

January 11, 2000. It was early afternoon and I was preparing for my second ecology class of the day. The weather had been particularly odd the last few days—warm, windy, stormy, reminiscent more of mid-March than January. The flu bug had hit the school hard and many students were in classes with flu symptoms, reluctant to stay home this last week of classes before semester exams. Needless to say, many of them were barely able to think, let alone stay awake for afternoon classes. The ecology students and I were working this day on the last three chapters of Rachel Carson's *Silent Spring*, where she introduces the reader to research in biological insect controls. I could already hear my mostly unhealthy students groaning at the thought of it. The students began entering the classroom, Carson books in hand.

We began the reading and discussion, and most students tried valiantly to concentrate. We were discussing insects' ability to resist pesticides, and the hikers and campers in the group were talking about their problems with horseflies and DEET. And then a miracle took place.

The monstrous rainstorm that had been going on for several hours ceased. High winds were rapidly moving the nimbus clouds, and all of a sudden a beautiful, clear rainbow appeared in the sky and seemed to be going into the lake. One of the students spotted the rainbow, and like automatons we all rose from our seats and went to the windows. The rainbow lasted about five minutes, and then it melded into the gray, heavy skies. No one said a thing, except for an occasional "Wow!" After it disappeared, I sat back down, and slowly everyone else got back to his seat, and the class continued where we had left off. It was one of those golden moments one cannot fully explain—sixteen people pausing from their daily routine to participate in a natural miracle.

I tell this story not only to relate a wonderful incident but also to preface this chapter on the subsequent years of the ecology curriculum experiment with an anecdote that points up the synergy of the ecology classes at this point in the school year. How we got to this point and how that pathway differs in small ways from the first year's path is the gist of this chapter.

Similarities to and Differences from the First Year's Group

October 15, 1999. I had introduced the Environmental Autobiography project and was beginning to collect the first rough drafts of the regression stage. The introduction of the project went pretty much as last year; students willing to participate but shocked to think I would be expecting thirty or more pages from each of them.

This second year I had two sections of ecology, the same as the previous year, but this time I had fifteen students in each section, as compared with the ten each section the year before. The two sections this year were considerably different from one another. The Ecology 1 group was concrete; they were not particularly interested in "what ifs." They wanted data, clear-cut assignments from me, discussion that didn't make them have to take any kind of personal risk. The Ecology 2 group was much more abstract. They were constantly pushing the envelope, asking more and more "could this possibly" and "can we try this next." They wanted the freedom to pull a discussion in any direction they chose and most would write lengthy responses in their essays and observations. I was determined to run both classes in the same way, mainly because of my desire to keep my involvement with the class similar to last year's, so I could make comparisons between both years' data. I am not so sure that I made the best choice pedagogically. I was prepared, though, to have very different results in the first stage rough drafts for the EA.

There was another aspect to this year's classes that was quite different from those of the year before: their general unwillingness to share any of their writing. Where last year's group didn't really clam up until the third stage, this year's group never opened up. About the only reading of the EAs that was done this year came from me. Once I realized that no one wanted to read his own work, I began asking certain students before class if I could read excerpts from their writing. No one refused me, and I think most were pleased that I asked them, but despite continual good, positive feedback from their peers, throughout the project I could not get anyone to willingly read his own work. Last year, once I kicked off a stage, it was the students themselves who almost daily reintroduced the project by wanting to read something they had written. Students this year were missing something by being so unwilling to put themselves out there and take a risk. For example, one day in early November we were analyzing our graphs from a human population study we did and comparing means, medians, and modes across the three cohorts we had decided to use. Clint, after class, started telling me how, while we were having that discussion, he was imagining himself in his eighties, what he would be like, how healthy he would be, and so on. I so wished he had brought that up in class, because I think it would have made an important connection for many students, in the throes of their progression stage. We could have discussed further the differences in male and female longevity that we were clearly seeing in our graphs, but that I couldn't get the students to discuss more than just a bit.

In this year's group of students there were several students who grew up in countries other than America. Hans was an exchange student from Germany, Brendan grew up in Japan and returned there frequently for visits, Vincent was born in Seoul, Korea, and Cody had recently come from England. These four students all happened to be in the Ecology 2 group, and this added a unique dimension to the project. The rest of the class was interested in hearing those students' EA excerpts and learned something about what it was like environmentally in a different country.

I continued to write written responses of some length in everyone's folder each time I read something new. After our first one-on-one meeting for the rough draft of Stage One, I decided to use only written responses and the occasional personal conference for feedback, as the additional ten students really cut into classroom time when I was trying to give everyone at least fifteen minutes for a conference on his writing. Some students still sought me out at breaks and before class to give me new additions and I would try to return those that day with written responses to what they had given me. I tried with each of these responses to always give some direction for the next writing, and generally students were glad for those ideas.

The pace of the writing was somewhat different. While the first stage provoked many exchanges of ideas and was easily written by everyone, there were at least five students still writing stories in their regressions as late as the second week in January. Many students had difficulty with Stage Two, more so than last year, and I didn't seem to be able to help them as much with the tools I had used last year—the Life List and meditation. Several students were only beginning Stage Three in January, a month later than they should have been. The seniors in particular were overusing their need to write college essays as an excuse for their paucity of new work, but I do have to add that once we were back in school in January, they did manage to reconnect with the project and turn out some prodigious work. I have tried every year to be flexible about the timing of the stages, but from the beginning the students knew the project had to come to closure by the end of the first semester.

The third, fourth, and fifth years were an amalgam of the first two. I continued to have about fifteen students per class, and one year I had three sections of Ecology instead of two. The EA has followed the flow of the first year more times than that of the second year. Generally the students have been quite up-front about reading their writings in class, but I also have been getting better at getting them to share their work. Stage Two has continued to be the most difficult of the four stages for the students, and on the whole, the Life List has been helpful. Some years the students are more open to meditation than others. September 11, 2001, was, of course, a horrifying experience for my students, and much of the writing that fall and winter connected in some way with that nightmare.

Narratives from Subsequent Years' EAs

From Clint's synthesis:

> When I first heard that we had this huge project due in four months, I thought it was going to be a problem. I was terribly wrong. This has been one of the most enjoyable papers that I have ever written. Everybody that I ask loves it too. I have fun thinking about my childhood and reliving those experiences. I will keep these papers for the rest of my life, so when I'm old and gray, I can look back on some of the best moments of my life.

Clint came to the class an avid environmentalist already. He had grown up in a family that loved the outdoors. He camped, hiked extensively, went fly-fishing and trapshooting, skied, and spelunked. When his parents built the home they live in now, the area was pristine and sparsely populated. Today, ten years later, the area has suffered suburban sprawl, and Clint and his father had gotten politically involved in some of the environmental issues of runoff, erosion, water pollution, and megahomes. Clint wrote in his analysis:

> When I am older and settled down with my family, I want to do some county work. I would love to be on the zoning board. I have seen how developers can strong-arm the zoning committee into letting them get away with lower acreage requirements. They go to court and plead hardship that they can't make any money with the zoning requirements, then the board gives in to the developers' whining.

Clint decided that the best way to celebrate the millennium would be outdoors. He arranged to hike to the top of Woodhull Mountain in New York State with seven of his friends. After a grueling three-hour hike through two feet of snow, they reached the summit just in time for the sunset:

> I looked at the mountains and the sunset like it was the first time that I ever saw it. We saw the sun set on the last day of the millennium and it was more of a spiritual thing than anything. I felt so close to my friends and to the landscape that I thought I could stay there forever. We stayed until the sun inched below the horizon and then the cold forced us back down the hill.

On our first day back at school from Christmas break, Clint immediately shared this story with us, intent on letting everyone there know that he was thinking about ecology during those millennial moments. The class was silent for a minute afterwards, imagining themselves on the mountain with him. As the months of school passed, I watched Clint become more and more aggressive about his ideas for his community and for our own school community concerning land use. One can see this progression visually in his EA.

Jared never did get to Stage Three and Stage Four in his EA. It was a shame, as he began the project with a large burst of energy and wrote a wonderful Stage One, and eventually, after much prodding, did a fine job with Stage Two. But I could never get him to spend enough time during the middle two months to prevent him from having to do Herculean work at the end. By the time the deadline was here, he simply ran out of time to finish. Jared was an intelligent, enthusiastic student with some attention deficit problems. It was difficult for him to stay on task for any significant length of time if he was chained to a chair and a computer or a desk and a classroom. The year before, as a sophomore, Jared got involved with one of that year's ecology students on that student's second semester ecology project. Every spare moment he had in school, he spent helping this individual work on cleaning up a section of the campus that had been spoiled by a small oil spill and a sizeable amount of dumping. By the time they were finished, the two of them had created a beautiful, clean section of the stream that runs through the woods and had added a small waterfall and two benches. When Jared began school this year, he was determined to take Ecology so that he could get involved in a similar project once again. I don't think he realized that that type of project would not even start up until second semester, and even then, would not be a large part of his weekly classes. But he participated good-naturedly in the class, even though it didn't turn out to be exactly as he had anticipated. His father was a surgeon who did a sizeable amount of charitable work, and Jared often accompanied his father into the operating room for those procedures. He had developed a sensitivity for the poor and was often the one voice in discussions that kept reminding his peers that not everyone was as fortunate as they. In one discussion we had on Darwin and his use of Thomas Malthus's mathematical ideas about population growth, the class was beginning to think that maybe Malthus's ideas about non-government intervention into people's social welfare was not a bad idea. It was Jared who, after listening quietly for several minutes to his classmates' descriptions of the "average person on welfare," reminded them that a heartless, mechanistic approach to humankind was one of the reasons the class was having so much trouble accepting Darwin's ideas of evolution.

Jared was sometimes seen at school as a "hippie." He had difficulty adhering to the dress code, he thrived on figuring out ways to break rules, and he had almost no conception of time. He wrote in his progression:

> I see myself owning a ranch in central Wyoming . . . making my living just enough to support my wife Rachel, and my children Lilly, Chase and Audry. We each have our own horses, and we ride through the backcountry, going on extended camping trips through the Gros Ventre and Beartooth wilderness. My children and I go hunting occasionally for venison, while my wife, a vegetarian, would eat from the garden we'd grow. . . . My children would be home taught, learning how to survive both intellectually and physically in the tough surroundings. . . . We'd take trips all across the country, my family and me, seeing everything worth seeing from the Rio Grande to the rainforest of the Olympic peninsula. We'd see everything, all the while recognizing the rules of environmental impact, i.e. leave only footprints, take only memories, and when you can, don't leave the footprints. . . . I wouldn't be making big bucks, but I'd be supporting my family and doing what I loved and being where I loved with the people I love.[1]

From Cameron's synthesis:

There have been things that I picked up from my ecology class. I came into [sic] thinking that this would be my senior year science class, and that I had bypassed taking physics, which is a course that I do not want to take. During our first week, I learned about what the course was going to be about and I was very disappointed. The thought of nature and its importance was not a common one, and needless to say, I did not look forward to Ecology class. However, I have learned so much, and most of it has not even been with the things we do in class. Some of the insights that the class has been taking are priceless to me. Just talking about things that go on and their relevance to us have been the things that I enjoy most. It puts everything into perspective.

---

[1] Jared today is heavily involved in environmental studies in college. Sustainable agriculture has particularly attracted him, and he has already experimented with several garden plots on his campus. He also has plans to travel to Central America and help with agriculture projects there.

> Some things that we have talked about were things that I would never had paid any attention to before. The WTO riots and that woman who stayed in the tree for two years are just two of the things that will stick with me. I realize the importance of everything around me. I no longer take anything for granted. I know what people have sacrificed for me, and I greatly appreciate it all.
>
> For example, I have taken a larger step into preserving the environment. Unfortunately, I do not know where to take it, so I cannot actually do anything about it. It is, however, all in my head and hopefully, one day, I will be able to put some of it into action. I do not know where to go though. It all seems so insignificant. I could work my whole life at cleaning up the earth, but it would not make a bit of difference. There is just too big a fight out there. The World vs. me is not a battle that I want to fight. Until an opportunity presents itself that will allow me to take action and actually make a difference, my dreams and aspirations for a better place to live will have to lie dormant in my head.

I had taught Cameron previously and knew him to be a thoughtful, hardworking student, quiet, somewhat disillusioned with school, and prone to periods of depression. The year before, I could already see him starting to get political as we would look at bioethical issues, but this year I was excited to see him become even more forceful. He had already come up with a political issue to explore in his second semester EE (Ecological Experiment) project—the problem of nuclear waste and what it means specifically for our area, as this area is supposed to be a major conduit for the transportation of nuclear waste from states north of us to a dumping station in Nevada. Throughout his EA, one is cognizant of Cameron's coming to some turning point in his life, where awareness of what he has done in the past has made him think deeply about himself in the future. He ends his paper:

> If I am sitting in my room, 50 years from now, reading my environmental autobiography, and I still have not acted, then I will be extremely disappointed in myself. Words cannot even begin to describe it. I only hope that I will do something with my life. With all of the advantages that I have been given, it would be a shame to not use them. I need to act.

From Anthony's synthesis:

> Overall, this was a very exciting, fun, interesting project, and it really did help bring back some lost memories. Though, throughout the duration of the writing of my autobiography, I never really discovered anything surprising about myself. I can probably contribute [sic] this lack of surprise about the development of the nature-self, to the fact that I have not changed drastically over these past seventeen years. I have remained the same little kid that I was many years ago, only my body has developed more. In fact, the only thing that I was mildly surprised about, was the fact that I really have not strayed too far away from that little boy who liked to always be outside, and just merge to become one with nature. Writing this paper has really

opened my eyes to the way in which I would like to live the rest of my life. I am not going to waste my life away by sitting in some dark, clutter-filled room watching TV, rather, I am going to be out there making things happen. . . . I think being in school for such a long amount of time is taking away from my real education. I wholeheartedly believe that the most valuable knowledge that one can hope to acquire is not one that stems from years and years in some yuppie institution, more wisely, they should look to find their place in the *real* institution of the world, nature. To sum it all up, I turn to one of the most radical thinkers and writers of all time, Henry David Thoreau: "Don't let school interfere with your education."

I had taught Anthony previously, had worked with his father, and have kept in close contact with him for several years. I agree with his assessment of his EA—he had changed relatively little. He was still the same idealistic, empathic, humorous person. But I did see one major change in him. Anthony was a jock. His life revolved around sports, and he played an interscholastic sport year-round. He often joked about his anti-intellectualism and called himself too stupid to be able to understand abstract concepts. Yet in this project he had thought deeply and seen that in his writing he could work through an issue and argue within himself. Because he was humorous, he often did these debates in an amusing way as he wrestled with issues of urban sprawl, animal liberation, and the value of an education. Since early November he had taken on a more forceful role in class discussions and was much more willing to go against mainstream thought, to play devil's advocate, or to remind a classmate that he was forgetting something in his argument. Because he played a leadership role in the school, he easily controlled a discussion with his common sense and his idealism. The EA, too, allowed him to explore his sensitivity and acknowledge that that was just as integral a part of himself as his highly developed cerebellum.

From his regression:

My mother helped me recall a time when she was very worried because my grandmother and I went on one of these walks in the middle of the winter. [He has written of his love of long hikes since he was two years old, where he calls himself "the greatest pathfinder of all time."] The two of us had set out for a mid-afternoon adventure. My mom became worried when it was well past dinnertime, and we had not returned home yet. Soon, my accomplice and I were walking up the driveway to our house, laughing and talking about the walk that we had just returned from. I accounted how I never seemed to be worried or in any sort of hurry when I was walking through my territory. I always seemed to so engulf [sic] in the surrounding environment that nothing else mattered. Every time, I had some new experience that would only further my appreciation for the world around me. I even thought of going on these walks as escaping from the pressures of being a young child. However, it may seem hard to believe that I actually had some hardships as a young child, I did. Even if I didn't have anything else to really worry about, I felt as though I was going to my own world, my own "Wonder land."[2]

Anthony would later move to a very different environment and it would take a while for him to feel as comfortable with his surroundings again. I feel he used his EA as a means of helping himself understand that transitional phase of his life. He also had numerous stories about his relationships with animals, some of them very funny. He wanted to learn more about animal behavior, but his method was strictly trial and error and it was only in this last year, with the help of his girlfriend, that he realized one could actually study some things about these different animals' behaviors and prevent tragedies from happening. He and his girlfriend spent much time in the local parks doing long-distance running. Again, from his regression:

> Have you ever noticed that there is a scent that goes along with summer? I am still not sure exactly what it is, but I sure do like it. It seems to be a combination of freshly cut grass, blossoming flowers, morning dew, suntan lotion, newly hatched eggs of a bird species, sweat from ambitious little athletes, etc. I cannot really describe it. While walking down the trail that leads to the athletic fields, my eye used to always catch something new in the wooded area that surrounded me. Though it might have just been a glance in real time, it seemed like it had been permanently etched into my brain. Vivid images of squirrels chasing after each other, a group of fallen branches that seem to make the outline of a particular image, the imprint of a deer hoof right next to one of my feet, and that is only the first thirty feet of the trail.

In another section of his regression, he wrote about seeing birdhouses on the school campus over ten years ago, birdhouses that have since been destroyed by weather or vandalism. He wrote that seeing those birdhouses on the campus made him happy to know the school would wish to celebrate the birds in that way. He decided to reconstruct some of those houses for his second semester EE project. Without the EA, I doubt that he would have remembered those birdhouses and then made the connection between those and how he could do something himself to help preserve an ecosystem.

From Kyle's synthesis:

> I have learned many things about myself as I have been writing this autobiography. Yeah, I am sure many people are saying the same thing and it does seem like it's a kiss-up line. I am serious, though. I have known for awhile that I do love the outdoors, but when I looked back and even when I was writing the second stage, I saw

---

[2] Anthony's parents were going through a rocky point in their marriage in these years and would subsequently divorce.

myself usually in one type of environment. That environment was out in the cold, high above, and where people aren't around.

This was the second year I was teaching Kyle. He was an energetic, almost rambunctious senior, who always seemed to be high on life. He was very much a social being, and yet he wrote extensively in all four stages of his paper about his need to be alone, away from noise and other people, to think. He took one of the most creative approaches to Stage Two, writing vignettes of his future life in dialogue form—conversations between him and some other person, in which the reader gets not only a sense of what is happening to Kyle but also what Kyle is thinking about the new experience. An example:

> "I can see it, I can see it happening. Don't you picture it?!" I said to my partner. I was in the office that day looking over some sketches and some pictures of the site. I really wanted this idea to happen, I thought it was perfect for that piece of property. I just saw that drawing working, but of course my partner disagreed. Oh, well, we will work it out somehow. Usually we take both our ideas and mix them together.

Just from this short vignette, one gets the sense of Kyle's creativity, enthusiasm, and democratic nature. Kyle had struggled mightily with ADD (attention deficit disorder) since second grade. He was matter-of-fact about it and saw the extra effort he had to put into his schoolwork as something he just had to do—no complaining, no trying to duck the issue. As he wrote the different stages of his paper, he would show me pages several times a week, asking for critique. He later enlarged this project for a creative writing class he was taking as an English elective and would continue to add to all his stages, especially his regression, through the second semester. He was keeping a daily journal. I had several faculty tell me that if anyone would have told them that Kyle would have been writing so extensively his senior year in high school they would not have believed it.

I am compelled to include a large section from Shawn's synthesis. In fact, his entire EA is so good, so insightful and well written, that it deserves much more attention than I can give it here. Shawn had already turned eighteen when we began this paper the previous September, and maybe that helps explain why he was able to take this project to almost another plane than the rest of the group. Something in this project pressed a bell for Shawn, and a distant, quiet, mechanistic student transformed before all our eyes.

> It is interesting to try to recall old memories. Often it seems in thinking about things in the past, that they must have happened to a different person. It seems unreal now

to have been the primary observer of the lightning strike near our cabin, for example. In some ways I don't really want to remember that that was a part of my life and in other respects it is nice to know that that is a part of me. Instances like that, that occurred when I was ten years old and younger or those that were too emotionally powerful at the time are relegated in my memories as though they are someone else's experiences. Even seeing pictures of past instances seem a little like seeing pictures of the history of someone else and truly understanding what happened to them.

A good example of this is the goat slaughtered to feed to the Komodo dragons on Komodo island. I remember the 3"x 5" Fuji color print that I took of the goat being killed more than the actual incident. I see the picture like someone else had taken it. The person that felt faint and went and sat down after seeing my friend the goat slaughtered seems to be someone else. [Shawn had been playing with this goat just prior to its killing, unaware of what was imminent.] This feeling of detachment is true of many memories. I can remember the details but they don't often portray me the way it seems I should have been at the time.

This project has forced me to dig up memories that I think about periodically, and actually put them on paper. I can relate much better to the boy who went tapping maple trees at a young age on his own and the one who spent a day playing with orangutans in Borneo than the one who almost got killed by lightning and by falling off a cliff when leaning on a rotting branch that broke. They are both the same person, me, and I would rather forget the instance of almost falling off the edge of the cliff since it seems to be something that happened to someone else. I mention this latter incident here but not in the regression because I did not want to remember it there. It seems unreal. I remember leaning on a log and leaning out over the edge of a dropoff, somewhere in the Holden Arboretum, having the branch break and then being rescued by my father. I remember I was wearing a red raincoat and the log was probably rotten and in retrospect the whole scenario seems stupid. At the time it seemed the raincoat made me invulnerable and safe and dry. I don't have any idea why it seemed that way. But then again, it seems everyone does stupid things at some time or other. I have never written or spoken about this incident before and this has probably helped give it a dreamlike quality. Even though my rational mind perceives this memory as real I would prefer to forget it and detach it from my history. I can almost convince myself that it was a dream and maybe it was, I hope it was.

At about the same time, about five years old, I had a carpet in my room with various square designs on it and at night I would sometimes dream that the squares opened up into bottomless pits. Sometimes they were bottomless and sometimes they were just very deep and witches were at the bottom torturing all the unlucky boys who stepped out of bed at night. The bed was safe because it was larger than the bottomless pits and the bedposts would fall into the pits but the bed could not because it spanned several pits. This probably seems perfectly irrelevant and ridiculous but it is about as real in my memory as almost falling off the edge of the ravine a hundred feet or so onto rocks below.

This has been interesting because it is the only exercise I have ever had where it seemed acceptable to ramble on about things not directly related. I can see more how my own memory works and find it a little scary how reality, unverified information and dreams coalesce into what is memory.

Shawn was a member of the Ecology 2 group (second year) and he often chose for himself the role of devil's advocate. As the months progressed, he learned how to energize a discussion. The first few weeks he would interrupt discussion with a verbal attack on whoever had been speaking and talk hurriedly and in a clipped manner, almost as if he were running out of breath. He learned to temper those inclinations in several ways. He now talked more slowly, apologized ahead of time for disagreeing vehemently with the previous speaker, and went on to present his case. He knew a great deal about hydrogen fuel cells, and the rest of the class would tease him about his ability to bring them into almost every discussion we had. He was learning to laugh at himself and to not take himself quite so seriously. He was torn between a life centered around computers and a Thoreau-type existence. His family owned a cabin in a remote area of northern Michigan, where at least once a year Shawn could retreat, to be without most of the amenities of modern life, including his beloved computers. He told the most wonderful story about himself when he was six years old. With very little help from his father and his uncle, he tapped into all the sugar maple trees on his property—a sizeable number—and produced his first of many batches of maple syrup.

From Harry's synthesis:

> At times I have been down. I remember how this summer I was depressed because I knew few people my age and I missed my family and friends in [our city]. The entire day was rainy and cloudy and that depressed me even more. I was sitting inside of my sister's apartment watching TV and I walked outside onto her deck. I looked out towards the west and my views changed. My depression drained out of my body. I looked out and saw the most gorgeous rainbow coated with small drops of drizzle. I laughed at myself for not appreciating the beauty I was surrounded by. Now when I get depressed I can think back to that moment and put myself back on that balcony at my sister's apartment. If I shut my eyes I can almost feel the same feelings I felt when I was there. Once again the pain and depression drains.

I have known Harry for the last five years and was now teaching him for the second time. Five years ago his parents initiated a slow, messy divorce and I have watched Harry suffer through that transition. Even before that, Harry had some difficulty with controlling his temper, but over the last five years, this has become a major problem for him. He talked about that very openly throughout his EA. Harry was one of the funniest people I knew. He had a great sense of humor, and when he smiled his entire face became that smile. Yet most of his EA dealt either with his temper or other periods of depression. From his regression:

> Remembering my past has brought smiles to my face, but along with those smiles came frowns of sadness and regret. I have many memories all coming back to me from the time when I was young till now. . . . I look at old pictures and see things that bring back memories. Dealing with my past causes me to feel things I have not felt in a long time. I have many regrets and much confusion in some memories. I wonder why I did that or what caused me to do what I did. I wonder if it could have been different.

I would have predicted that Harry would sign up for Ecology class for his senior year. I have always known him to be very interested in the environment and to have strong opinions about ecological issues. For example, he didn't understand why anyone should have more than one child, even though that would mean he would never have been born. He and Shawn had been battling adversaries since the first week of class, and I had to remind both of them on several occasions to debate within the confines of orderly discussion. He once read this section from his progression in class, and Shawn pounced all over it:

> I look into what I hope my future is and I don't see really what I am doing but more what I hope I am not doing. I have many things that I hope I stay away from. My number one rule is to make sure I succeed and avoid failure at all costs. Success has nothing to do with how much money I make or what kind of car I drive. It has nothing to do with my office size, if I even have one, or how many people are below me in a working sense. Success for me means happiness. My fantasy is to raise my kids right. I hope I do all for my wife that I can possibly do. I want to take care of my parents and support them the same way they have supported me. My father told me two things that I will never forget. One is that if you are going to do a job then do it right, otherwise don't do it at all. The second thing is that it does not matter what you do in life as long as you do it right. . . . I will follow my dreams and aspirations. They will lead me to happiness and happiness in life is all any person will ever need.

Harry seemed not to care that Shawn and others in the class thought he was being too idealistic and too general. In several written notes to him, I asked him to define what he meant by happiness, but he never did. For now, what was important to him was that he was affirming his father's advice. I think for the first time, with the help of the EA, Harry was seeing that he could control some of the things that happened to him, while also learning to control his temper. And for now, that was lesson enough.

From Brendan's synthesis:

> Flipping back through the pages, reading all the things I have written, I realized that they weren't just words on a paper, and they were not even just words about me:

they were words that showed the kind of person I am, and from those words, I got a pretty good sense about who I really am, for the words were my own, they were a product of my own mind. That is what I learned is so special about that kind of writing, that along with writing about yourself, about your life and your thoughts and your ideas and your desires, you're also getting a good sense of yourself by not just the words you write and what information you give, but about *how* you write, about your style, which gives so much more information than "I like the ocean" or "I wanna be an English teacher when I grow up." And the best part about my realization of who I am was that I was very happy with it. I am very happy to be the way I am.

Brendan was disarming. When I first started teaching him, he told me one day he was the dumbest Japanese person I would ever meet—"I'm just a dumb Jap," he said. He called himself that on numerous other occasions, especially when we were doing something mathematical in the classroom. He would also stop a discussion cold in class by asking me something very personal. Once when we were talking about the role of pheromones in insect control, he asked me why human semen smells the way it does. During a class when we were looking at Kammerer's experiments with toads, wherein he misrepresented his results in order to win acclaim for an attempt to prove Lamarckism, Brendan was dumbfounded to think a scientist could lose his job for something like that.

Yet if you could read his EA in its entirety you would be impressed with his language skills and his ability to describe an environment down to the minutest details. He wrote one story in his regression about a recent trip to Columbus by himself. He decided, once there, that he would be too bored driving the same route back alone, so he went to the library and checked out where there were covered bridges he could visit on the drive back. This seventeen-year-old boy took hours longer to return back home in order to visit several covered bridges, and then he described them in loving detail in his story.

In his synthesis, he writes about "the integration of your body and your physical environment":

> What do I mean by that? I am talking about the way your body, your mind, works with the environment around it, the people, the places, the things you see, touch, hear, feel, smell, and taste. It is something that happens all the time to people, but since it happens so much, it is often taken for granted. Words cannot express the greatness of power that your physical environment has on your body, how it may even completely take over your body. Imagine something: put yourself in a place that is special to you like no other place in the world, a place where you feel completely at peace and no worries can ever enter your mind. It may be a beach, a park, a room, a forest, a car, anywhere. Imagine yourself in this spot, and allow yourself to feel everything without thinking about it. To me, that place is in Japan, the whole walk from my grandmother's house to my school a 10 minute walk away. When I walk, I see the narrow streets that my friends and I used to play freeze tag in, the

house I have spent my summers in, I hear the trickling of the little stream behind my house, I hear the crowing of the caged rooster in my neighbor's yard, I smell the fresh fish coming from the outdoor fish market up the street, I hear the train rumbling under the bridge I cross, I see little old women riding their motor scooters to the grocery store, and many, many more things.

Brendan told a wonderful story about how his grandfather took him up Mount Fuji for the first time, and what a magical moment it was when they finally reached the top, after a fog-filled journey, to have the fog lift and see what he perceived as Tokyo in the distance. His regression and progression were full of wonderful stories. Toward the end of his sophomore year, Brendan's father died, after a short, horrible battle with cancer. Last school year, he was bereft, but it has been amazing to see the difference in him this year. Even though he didn't mention his father too many times in his paper, I think the EA helped him focus on new adventures for his life and allowed him to move on. He ended his paper with a challenge to the reader:

My message is simple: for the past 35 pages I have thought about my life and learned to appreciate it more fully and learned about how to make it better. I know no other way of saying it, but it feels good to have done that, and now I wish to pass on that task to you, the reader. Do the same as I have done. Look at yourself. Write about yourself. Get to know yourself better, and get to learn what you really desire out of your life, and then go out and get it. It is an amazing thing to hear about yourself, for though you are with yourself more than you ever are and ever will be with anyone else, sometimes it is yourself that you know the least about. Be true. Be honest. And enjoy.

From Byron's synthesis:

I noticed that my childhood experiences of learning lessons on my own, without ever being told what was right or wrong, has formed me into someone who for the most part now makes the right choices. I'm glad, as I said earlier in this paper, that I have had a chance to do this. I know someday I'll read it and see how accurate I am or am not.

This was the third class I had taught Byron. He was a junior and still quite a concrete thinker. He had difficulties with spelling and grammar and generally disliked writing, I think mainly because his papers so often had more red marks on them than marks of his own. He really got involved in this project, though. He was the only person either year to argue with me at the end about keeping the paper in four distinct chapters, the four stages. He wanted to meld his paper into one, and he began his paper with the sentence, "It's possible that the hardest thing about this paper is figuring out what order to put everything in." I learned many new things about him from the paper: that he loved to clean house and cook, that he saw himself as owning a restaurant in the future, that he had such a strong sense of what is right and wrong for him that he often couldn't sleep when he thought he had done something he shouldn't. He wrote a long story about putting his initials in wet cement on a neighbor's driveway and spending literally months of anguish until he finally confessed to his father and then to his neighbor. He tried hard to bring the environment into his stories, but he kept getting more interested in the connections with his developing sense of morality. My favorite story is this one:

> This next memory is with my dad. It's not really a memory but I guess it is. It's a perfect lead-in to the present, though, because it has happened since I was about 3 and it still happens now. This might sound a little weird but it's cool. Every Saturday (that we're both in town) we go out for breakfast. It's not a normal breakfast, though, it's different. I used to alternate with my brother when he used to live at home (now at college), but we used to go this small restaurant owned by an older woman. She knew us and we knew her. Of course we never knew each other except for through that restaurant, we knew each other by name.
>
> Anyway, our Saturday morning breakfasts are different than most traditions, because we write letters to my grandpa (and grandma when she was still alive) who live in NJ. I didn't and still don't get to see them very often, maybe once or twice a year, but we communicated every week just about. Before I could write I went and dictated to my dad what I wanted to say to grandma and grandpa.
>
> It has really become a cool tradition. Something that both of us are going to miss (when I leave for college). I think we both look forward to it really. It's a good time to talk, write, and just be together. After the small restaurant closed down or went out of business we have not found another place to go regularly, so we started just going to all different places. Everywhere we go we evaluate. They don't know we're evaluating them, but we are. And we take notes sometimes if there is something we really like and write it to my grandpa. It's a funny thing but I look forward to it every Saturday, so, Dad, if you ever read this, thanks for our Saturday morning memories and I hope we make plenty more.

He did write quite a few stories about the environment, but in every one of them he gets hurt and then he calls one of his parents and they take care of him without admonishing him and he ends up admonishing himself and appreciating his parents' response. As he gets older, the risks he takes are more serious, and the last, quite lengthy story he wrote talks about his experimenting with drinking (it starts outdoors!) and goes on to his learning quickly that is not for him and his parents' response, which again he appreciates. He melds these experiences into the future and before you know it, we see him and his own children having Saturday breakfast together:

> My favorite days are Saturdays. Saturday mornings are the only mornings that Gustav's [his restaurant] is open for breakfast. We have Saturday brunch. It's a whole family ordeal. Children are free, but I'm never there, because I go out to breakfast like my father and I used to do, only with my kids, and we write letters to my par-

ents. It's weird, my parents live just 45 minutes away, but the tradition just stuck with me. And see, we write every Saturday. I try to get Tai [his wife] to join in with this letter writing, but I guess I can understand that it doesn't quite mean that much to her, because she never did it when she was younger. My kids usually just tell one of us what they want to say to their grandparents and we write it for them. Our five-year-old can now write his name, so he signs the letters.

Timothy Leary, an expert on temperament before he dropped out of academia, was prone to ask people about the circumstances of their lives when they were seventeen. He found that our basic attitudes about ourselves and the world tend to be meshed into "the rapidly setting concrete of our nervous systems around that age," and usually change very little thereafter (Gallagher 1993, 166.) This idea is prevalent in so many of the EAs but never so clearly delineated as in Byron's paper.

From Zach's synthesis:

When I am out enjoying nature, and interacting with my environment, the experience that I love most is the feeling of peacefulness, freedom and escape that I often feel. All of my most memorable experiences with my environment have to do with these feelings. These feelings also only come when I am in a natural or preserved part of my world. I don't feel this when biking around, or when playing football, or even while going to the New Jersey shore. These feelings only come out when it is just myself and nature, the way it would've been 1000 years ago if I was there at that time. These feelings are what I strive for when interacting with my environment. They bring an indescribable feeling to me and my entire body. They bring feelings of joy, peacefulness, and freedom that you don't feel in your neighborhood, and especially while watching TV. They provide a wonderful escape from everyday life, and when experiencing this I feel as though I'm being cleansed of all of man's wrongdoings and of all my impurity. That is the best I can do to describe it, but I recommend that everyone try and feel that once in their lives, because it truly changed me as a person and my perspective on life.

Most importantly, I learned how nature has helped shape me into the person I am today, and without my experiences I wouldn't be the same. My experiences have made me more environmentally cautious and aware. I feel they have also made me a more understanding and well-rounded person. Due to all my exposure, I feel I understand more about the world around me and people around me. . . . Nature has provided me with what makes me unique and different. It's what separates me apart from anyone else.

When I decided to try the *currere* approach with my Ecology students, I was mainly interested in finding something to help them connect with the environment. What Zach has written is what I would hope for all my students—that on some level, they saw that only within nature could they have certain kinds of experiences, experiences that were essential for their well-being. If those environments were no longer as accessible to them, they would be missing something that could not be replaced.

Prior to beginning the EA, I knew Zach had had a unique camping experience in Colorado the past summer. He and I had talked about how great it was that he would have a chance to write about this trip in his EA, something he had wanted to journalize anyway. The months passed and Zach had written voluminously in his regression and progression stages, but still no mention of Colorado. "Zach," I said, "don't forget to save a block of time so you can write about that trip." "Oh, yes," he assured me, "I definitely will." Christmas break came and went, and I thought to myself he would probably use some of that time to write that section. A week before the final paper was due, Zach still had not written one word about Colorado. When I met with him to admonish him to get going, he confessed that he didn't know how to start. He reminded me of Carson the first year, trying to write about his last canoeing trip where he became involved with the native Americans he met on the journey. Carson couldn't write about that either, and he ended up writing very little about it, saying that no one could possibly understand what that experience meant for him. I could respect Carson's decision then, but I felt he was making a mistake not working through those feelings on paper. I didn't want the same thing to happen to Zach. I told him of Carson's frustration, and I could see he was strongly identifying with it. I crossed my fingers behind my back, reminded him of the deadline, and encouraged him to try one final time. I found out later that that night when he got home, he put a piece of white rock he found rock-climbing on that trip next to his computer and began to type. He ended up with pages and pages of story.

I taught Zach for two years, and I have watched him develop from a rather flighty, fun-seeking boy into a thoughtful, empathic, fun-seeking teen. When he enters a discussion, he is deliberate in his choice of words, and everyone stops to listen intently. He is not strident or dogmatic but instead gives a reasoned response. I have been talking to him about considering a career in environmental law.

From Edmund's synthesis:

> I have learned that my interactions with nature have greatly changed me as a person. I remember making my fort in my backyard using old planks, then my brother and I named it Fort Seneca. I don't know why we did, but to this day I remember that. I remember leaving it up all winter and having a huge snow war there. To this day I still am building things with my hands, whether it be a stereo speaker box for my car, a piece of furniture for my bedroom, or just a piece of sculpture for my art portfolio. I think that I enjoy working with what comes from the earth.

> Just this year I have begun a small flower garden on my windowsill in small Dixie cups. I think it is very relaxing to look up and see flowers drooping over your head as the sunlight passes through them. It's not as good as the leaves at my old house but it will have to do. I think that as I grow like my flower I will bloom and go towards the light, meaning greatness. Or at least I'm going to try.

The previous three years of high school had not been pleasant years for Edmund, and if one could have seen him in the hallways then, one would never imagine that person with a row of Dixie cups on his windowsill. He talks in his autobiography about his breaking rules at school, his unhappiness with his peers and the school, and his discontent with life in general. Then something began to fall into place for Edmund at the beginning of this, his senior year. I think in a small way the writing of the EA helped him analyze his past feelings and think seriously about his future.

Edmund had many interests outside of school—working on cars, racing stock cars, and working for a veterinarian at an animal hospital. His parents divorced when he was in primary school and he talked about how different adults had influenced his involvement in these various activities. And he talked about how he envisioned his life returning to what it was before the divorce:

> I loved my old house. I can actually remember how my old house used to smell at different times of the year. I remember that when I would walk into our sunroom I could make out the smell of the cedar on the walls. . . . I loved the scents. I remember the way I used to chill all day on our large, green leather couch and watch television . . . . I remember all of the events and gatherings that took place in this house. I loved the way our kitchen smelled on Thanksgiving. . . I loved looking out of every window and seeing a different sight every time. I loved my lake. I loved sledding with my siblings on its hill. These are the things that will stay with me forever and I will never, ever forget them.

Edmund does a particularly thorough job with his progression. He writes several scenarios for himself, one isolated on an island where he cares for animals, one in a midwestern town where he races cars for a living, and one in a large city where he works in an office and does sculpture as a hobby. In each of these, he goes into great detail as to how he would live, from describing his habitat to building a daily schedule. In all of the stories, one can see his engineering bent and his need to be creative. A small excerpt from his Stage Two:

> If I were going to drop out of society, I would buy a boat and sail to a deserted island or a small-populated one. I would own a shelter, kind of a house on stilts. It would be very similar to the one in the movie Swiss Family Robinson. The house would be definitely in a tree and I would have to bring my animals with me. To enter the house you would have to pull a vine that would lower a ladder-type staircase

that would keep out any unwanted guests. Then the first room that you would enter would be the foyer. In here would be a mat for shoes made of seaweed. From here you could go to the right or to the left only. If you went right you would be in my kitchen, that would have running water that I could work on a system I made with a pump. The small Honda generator would create just enough energy to power my refrigerator. The table would be made of bamboo legs and hand woven bark to create the tabletop. Attached to this room would be my family or entertaining room. I would have my paintings and artworks posted on the walls, a recliner that I had brought with me would sit upon a Persian rug. And a picture of my car would sit in place of a television.

Apparently, so he could watch it and remember.

During the fourth year, two catastrophes hit the classroom: the nightmare of September 11 and the unexpected death of one of the ecology students on December 30. By the time the students were ready to begin the Fourth Stage in January, the synthesis, most of them were depressed, confused, and scared. One of my students had actually witnessed the student's death, and another had been that student's closest friend since early childhood. There were days when neither of these two boys could even get out of bed. The wrapping up of the EA took on special significance for everyone.

To commemorate that boy's death, some of the Ecology students and I built a trail on campus and named it for him. The planning and construction of the trail was rife with problems; the students' unhappiness was palpable throughout. We ended up with a beautiful trail, but I think each of us who worked on it can still see drops of our individual blood somewhere along the path. One of the fourth-year students wrote this poem in his EA:

> Sometimes,
> when there is a break in the momentum,
> in the speed,
> in the blind rushing,
> the clouds may part,
> and,
> if you look heavenwards
>
> for just that one instant,
> that one moment of clarity,
> things once invisible
>
> jump out of the shadows
> and reveal themselves to you.
>
> The secret language,
> hiding in plain sight, of the whispering trees –
>
> swaying and moaning,

> incanting the prayers
> and stories
>
> of lost leaves
> and limbs
> past.

Summary

At the beginning of this chapter, I quoted David Lusted's definition of pedagogy—that transformation of consciousness that takes place at the intersection of teacher, student, and the knowledge they produce together (1986, 3). During the four-month writing of the EA, the students were not in stasis. Most were applying to colleges, all were attending other classes and participating in sports, other learning was going on in the Ecology class, and they were living lives outside of school. These present experiences were concrete and connected to how they were writing their EAs, especially in Stages Three and Four. In the production of this autobiographical knowledge, the students and I intellectualized ecological principles more deeply and recognized that the development of our consciousness is in constant flux.

I would be remiss in ending this chapter if I did not mention how I had changed in this process after five years. In the second year, I was more comfortable with *currere* and began to deconstruct it myself. The first year I was so bent on following the method and making it clear to my students that I did not have the presence of mind to analyze the method in and of itself. I, of course, from the beginning was analyzing the method from the standpoint of how it was affecting my class, especially in light of the fact that I was surprised by the energy it generated. But as the second year came to an end, I was beginning to analyze *currere* for its own sake, not simply as an educational tool I was using to make my ecology class more meaningful. The following are some of the things I learned:

1. *Currere* needs to be protracted. The method cannot be rushed. Because we are learning that we are all part of a community—the community of the ecology class, the community of the school, the larger communities outside the campus—we need time to develop this relationship and to let the regression and progression stages in particular take hold.[3] One of the first-year students wrote the following in his EA:

---

[3] Janet Miller (1990) says that collaborative relationships between teachers and students need to be of "long duration in order to take into account the complex constraints of those who want to uncover as yet unrecognized forms of oppression" (153). All the more reason to devote enough time to *currere*.

I think the problem was that I tried to sit down and start writing immediately, figuring that things would just come to mind. But that was not the case as I soon found out. It took a little bit of time where I just sat and thought about my past. After I did that, it became a joy to sit and record events in my life. It opened me up to the environment and helped me form the foundation for what I come to expect in the future. . . . I have come to recognize that the environment is not just a place where I live, but rather it is a main factor in the shaping of myself and the events in my life.

2. The four steps need to be followed in sequence. Stages Three and Four mean nothing without the "data" of Stages One and Two.[4]
3. Regression needs to be stream-of-consciousness. There can be no limits put on the student's memories. What he wants to write about in his regression is just as important as what he doesn't want to write about. The student has to feel that no memory is unimportant. If later on the writer wants to exclude a memory he had previously written, Stage Four is the place to do that. One of the students in the fourth year wrote, "When you find yourself out in the woods you become very inventive. That's what I love about being outside; you're given so much to work with and you can do what you want with it." That is the attitude one must have with the writing, too.
4. Analysis cannot take place until Stage Three. This is really an outgrowth of the previous statement. Attempting to analyze one's regression as one is writing makes the writing artificial. The memories seem planned—almost as if they were written to appease the teacher.
5. A computer is a wonderful tool for *currere* because it allows the author at Stages Three and Four to reconfigure his paper—move sections around because of common themes, for example. The student will be much more willing to do this if he can simply cut and paste with his computer.
6. For *currere* to be effective, the teacher and classroom must be prepared, readied, for it (more on this in the next chapter).

I don't think I would want to do curriculum as autobiographical text without using *currere*. *Currere* gives order to the process and provides a context for the writer to imagine, create, and relive. For those students who have difficulty letting their thoughts run freely, it forces them to try it. I repeatedly saw students who were at first resistant to meditative picturing of themselves

---

[4] Connelly and Clandinin's work on autobiographical text exhibits this same conclusion. "[I]nstead of asking people at the outset to write a narrative we encourage them to write a chronology. We avoid asking people to begin by writing biographies and autobiographies for the same reason. People beginning to explore the writing of their own narrative, or that of another, often find the chronology to be a manageable task whereas the writing of a full-fledged autobiography or narrative, when one stresses plot, meaning, interpretation, and explanation, can be baffling and discouraging" (1990, 9).

in the future (or in the past, for that matter), finding great pleasure in that process once they felt they had to try it. Over and over again I found students who could remember farther and farther back into their pasts because they were reenergizing those synaptic links as they forced themselves to look at memories in the early stages of regression. The remembering became easier and easier. They became flooded with memories.

As I became more comfortable with *currere*, I found I was able to give better advice to the students, to write more helpful notes to them, and to listen to them with a more critical ear. In the fifth year, I asked each of the students to write a section in their analysis (Stage Three) about beauty. Has not beauty been a subconscious theme in your paper? I asked them. When you wrote about yourselves for Stage Two, were you not picturing yourself in some beautiful place or structure? This turned out to be an important part of the analysis. Almost all the students did a wonderful job trying to analyze their own concept of beauty.

The next chapter looks at what I did to prepare myself, my students, and my classroom for this unique experience.

Chapter VI
My Pedagogy

[T]he fact is that science can't tell us whether to be alarmed [about genetically engineered plants]. Why are we still in ignorance on this important issue? One reason is that the biologists who engineer genes into plants aren't trained to think about the whole organism and its habitat; they're molecular folk, lacking an ecological perspective."
—John Bendit, editor of *Technology Review*, the Massachusetts Institute of Technology "magazine of innovation," July/Aug 99, vol. 102, no. 4, p. 8, in response to a series of articles on the genetic engineering of plants)

The Context in Which I Designed the Curriculum

The school where I teach, a private all-boys college preparatory school, is fortunate in having over two hundred acres of surrounding property, much of it wooded. There is a large six-acre lake on the property that is fed by a side stream of the Chagrin River. In conjunction with changes in the science department for the school year 1998–99, I was asked to come up with a class that would make more than token use of this environment. In school years past, classes from kindergarten through grade ten used the lake and woods for field trips, one-day/one-class events where the students would walk the woods and gather materials and data. My task was to come up with a class that used the environment as an ongoing extension of the indoor classroom. I was excited about this challenge, as I have always wanted to teach ecology to eighteen-year-olds. I was given carte blanche to design this curriculum; all I had to be sure to do was follow the bell schedule.

As I began to design this curriculum, I was looking for an innovative way to teach ecology, really a different way to teach all sciences. We educators are teaching science the way it was taught seventy years ago, yet we are living in a dramatically changed world. Artificial intelligence pioneer Roger Schank of Northwestern University says that science education is still "about preparing for Harvard in 1892 and not for life in 2003" (*Wall Street Journal*, B1, 12/27/02). We are teaching a different child, I would argue a different child than even ten years ago. This is no longer a world where schooling prepares you for a particular job, where Horace Mann's message to labor, that school would provide them with skilled employees, is the same message. Schools must rethink their mission in light of this changing postmodern world. It is no longer enough for a scientist to know scientific method.

In fact, the world has changed so drastically since 1960 that science education research has to rethink its tools and goals in order to better understand this postmodern world. Annette Gough has developed four principles for science education research that she feels will help promote this understanding: (1) to recognize that science knowledge is partial, multiple, and contradictory; (2) to draw attention to the racism and gender blindness in science education; (3) to develop a willingness to listen to silenced voices and to provide opportunities for them to be heard; (4) to develop understandings of the stories of which we are a part and our abilities to deconstruct them (Gough, in Pinar 1998, 196). How can science educators begin to address these four issues in their classrooms? Can they politicize their teaching? Marion Namenwirth says that many scientists "firmly believe that as long as they are not conscious of any bias or political agenda, they are neutral and objective, when in fact they are only unconscious" (Namenwirth, as quoted in A. Gough 1998, 196).

First of all, we have to get students to like science. This idea of making science so difficult as to keep it somewhat elitist is wrong. The basic concepts of science can be understood by most of us. Science is incredibly interesting, and we as educators need to impart that interestingness to students. For example, isn't it amazing how all body systems work together, how many fail-safe systems are built into physiological structure, how the size and shape of molecules tell so much about their ability to function in certain ways? How does evolution work, and is it over? Are humans still evolving? If so, what does that mean? Can we direct our evolution? In the field of ecology, how is it that as we gain a basic knowledge of how ecosystems work and how fragile they are on one hand, we also see how resilient and tenacious they are on the other?

I personally feel our planet is in dire straits. We are coming dangerously close to the carrying capacity of the earth; by the year 2050 most demographers see world population between 11 and 12 billion people, almost double where we are at the turn of the century (McKibben 1998). We Americans are particularly adept at misusing water, air, and soil. We are convinced that our technological know-how will fix everything; the hegemony of "technology as improvement" is firmly entrenched in our psyches. One way, one small way, we can deal with these issues is better education about these issues in schools. That is why the curriculum of ecology has to change—to dramatize that the earth we are dealing with today is not the earth of seventy years ago, or even twenty years ago. Ecology textbooks have paragraphs galore on the raping of the earth and the pollution of lakes and streams, yet it often seems that it is just something else to read. Students generally don't make contact with the information in those books. I am convinced that unless we get students to care, to feel, that the Earth is in danger, the teaching of ecological principles will not transfer to living a life based on those ecological principles. That is how I came to this idea of *currere*.

When I came across Pinar's idea of *currere* while reading his *Autobiography, Politics, and Sexuality* (1994), I knew instinctively that it was an idea worth pursuing for this new class. The idea of dialectical interaction, of taking the learning philosophy that so interested me and putting it into a practice that called for, shouted for, real connectedness with my students, fit in so perfectly with what I knew was an essential component of the teaching of this subject: the need for intense, habit-changing involvement, a "religious experience," if you will. *Currere,* Pinar's method of curriculum, depends on an internal dialectic. "It is a call to examine one's response to a text, a response to an idea, response to a colleague, in ways which invite depth [of] understanding and transformation of that response" (Pinar 1994, 119). Pinar was downplaying the elitist, intellectualizing posture of so many educators in which they proselytize their students, attempting to convert them to the "timeless truths" they have so painstakingly "discovered." Pinar sees life as one continual experiment, teaching and student in constant flux. So I designed the EA as a tool to help my students begin to understand themselves. How have the spaces in which they lived influenced who they have become? How will they choose space in the future to help them live their lives more fully? Pinar seems to have developed *currere* as a method for the curriculum theorist; I wanted to see what would happen when eighteen-year-olds used the same autobiographical techniques but within the confines of issues of space.[1]

After I had experimented with how to use the method in a high school classroom, the idea of writing an autobiography based on place seemed like a possible, workable premise. I put together a syllabus for the class, plugged in time for the four steps of *currere,* and crossed my fingers.

Syllabus

There are certain ecological principles that must be understood in order to make any sense of the holistic nature of ecosystems. Students, in my experience, are not used to thinking of science as holistic; they think of science as an unconnected group of narrow ranges in which each topic is covered and where no overlapping is discussed. For example, photosynthesis is taught as a series of light and dark chemical and physical reactions that occur because certain conditions are met. The fact that a butterfly could go through a metamorphosis from larva to pupa to adult seems, to the student, to have no link to photosynthesis. Yet the butterfly is totally dependent on the process of photosynthesis for its energy.

---

[1] I would later find that it was impossible to keep the students within those confines once the project really got going.

The first thing I did in creating a syllabus for this class was to make a list of the ecological principles I thought vital to my students' immersion in ecology: closed systems vs. open systems, trophic levels, energy transfer, succession, mapping, population studies, evolution, photosynthesis/aerobic respiration, the First and Second Laws of Thermodynamics. I then looked at how those concepts could be introduced by linking them up with something that was going on outdoors on the campus. Since September is such an active month visually in northeastern ecosystems, we began by observing different niches and writing these observations down in a journal.[2] We then took one of these ecosystems and tried to come up with a trophic level diagram for that unit. From there we could look at the populations of each member of each level and see for ourselves how the numbers changed as we moved through the different levels. This brought us to discussions of energy transfer, where the energy began (through photosynthesis and the first law of thermodynamics), how the process lost efficiency as the energy passed from one organism to another (second law of thermodynamics, entropy), and the ideas of interdependency and holism began to take shape.

This approach would make much use of our outdoor space, but that alone would not make the students into ecologists. For that to happen, I had to find the key that would "get them," that would pull them into this study of ecology to the point where they wanted to know more and wanted to become involved in the preservation of our outdoor space. Many ecology textbooks stress the relevance of taking one ecosystem and studying it in detail, a class case study if you will (Shrader-Frechette and McCoy 1995, 198–279). This provides the student with a real ecology experience and allows him to study that case for a protracted period of time. I had plenty of ecosystems to use for such a case study, but I also knew the students had had prior experience with ecology and hadn't realized it at the time. That is when I began to bring in phenomenology and the study of lived experience. Finding Pinar's method of *currere* was the key for me; it "got me" and sucked me in enough to convince me that the semester-long risk would be worth it. Even if the method failed, I figured getting teenagers to think and write about themselves was a relevant exercise in preparing for the future. I did plan some back-up activities using the idea of autobiography in case the EA was not working. I worried needlessly. (See chapter 3 for the methodology I used for the EA.)

Because so much of ecology is controversial (Ulanowicz 1997, 1–10), I wanted critical literacy to be one of the main focuses of the class. Wendy Morgan gives a generalized definition of critical literacy in the beginning of her book *Critical Literacy in the Classroom* (1997). In the deconstruction of

---

[2] The students and I kept this journal throughout the school year, and as the year progressed I made more and more demands on them as to what they should be recording in these journals.

a text, critical literacy readers, she writes, "focus on the cultural and ideological assumptions that underwrite texts, they investigate the politics of representation, and they interrogate the inequitable, cultural positioning of speakers and readers within discourses" (1–2). The text is explored contextually — who wrote it, what biases the author may have had, when it was written, how persuasive the language is, who may have funded the research for the article, what interests were served by the writing of the article, how might the argument be represented within a different context (2–3). This type of writing, I feel, is important for the writing of the EA; it prepares them to critically analyze their own text in Stages Three and Four.

We read many articles on aspects of ecology, and then we discussed them critically (see discussion later in this chapter on the technique I used to teach students to read and write for critical literacy). The students were always required to do some kind of writing to go along with these readings, and the writings were often the jumping-off point for our discussions. I found that these writings were helpful with the EA, especially the analysis stage (Stage Three), when the students would try to look at their own autobiographies with the "seasoned" eye of the critical reader (see chapter 3 for a discussion of Stage Three).

The second semester evolved around an ecological experiment (EE) that each student designed, executed, and reported on. Some students paired off on this project, but I tried to get most students to work alone; I was interested in getting as many different projects into the indoor and outdoor classrooms as possible. The students were allowed to choose a topic. As students were finalizing their choice of topic I became aware of how often those topics dovetailed with what they had written in their EAs (more on this in chapter 12).

Because February can be a particularly bleak month in our area, I knew we would seldom get outside, so I decided to take that month for a study of evolution. I approached evolution as the study of a theory, as was so much of the previous work we had done (global warming, ozone depletion, pesticide effects, etc.). I then historicized it. I took Lamarck's idea of species change and contrasted it with Darwin's. We then took several evolutionary case studies and looked at them through Lamarck's lens and through Darwin's. I had the students write those arguments, and we would then use those writings to initiate discussion in class. Once I felt the students saw biologically how Lamarck could not be correct, I proceeded to go into greater detail as to how Darwinism and neo-Darwinism work.[3] We explored the difference between

---

[3] Although we did read a Stephen Jay Gould article on the persistence of Lamarckism even today. Gould's premise was that Lamarck's theory was appealing. It left humans in control

allopatric and sympatric evolution and made our own little "textbook" of evolutionary principles. Evolution is a difficult topic for any student, and a teacher must be prepared with many tangible examples the student can use to work through these complex abstract ideas. But once there is some understanding, it is fascinating to have discussions on the future of evolutionary process—"Where are we going from here?" Very quickly, the student lapses into Lamarckism and wants to control where we are headed. Where might human evolution go? Where might arthropod evolution go? How will the drastic changes we are seeing in ecosystems affect the principles of evolution? Will the evolutionary process be speeded up because of what is happening to the planet? The students enjoyed this type of questioning in large part, I think, because of their concentration on their own futures in Stage Two.

The months of April and May are, again, wonderful months to be outdoors. We continued our observations, got ecology experiments to the point where the students were ready to begin collecting data, and used indoor classroom time to study global ecological disasters. Up to this point I resisted much discussion about global issues, preferring to focus on issues in our own backyards, although we had spent considerable time on global warming and ozone depletion. The disasters I focused on were Bhopal, India, the Exxon Valdez oil spill, the Times Beach dioxin contamination, Love Canal, Chernobyl, and the Three Gorges Dam in China, a newly begun project that I personally felt was an ecological disaster.[4]

We read two books during the course: first semester, Rachel Carson's *Silent Spring* (1962); second semester, Bill McKibben's *The End of Nature* (1989). Since we were not using a textbook, a conscious decision of mine,[5] I wanted to have some reading that was protracted. Carson's book would be an excellent addition to the students' personal libraries, and McKibben's book was controversial. I was particularly looking for a book that would spark, I hoped, furious debate.

The EA became the core of this curriculum. It seemed to be driving all the work we did in the second semester and had driven most of the work in the

---

and discounted the idea of evolutionary "luck." The students and I also discussed how Lamarckism seems to drive evolutionary psychology today.

[4] I was adamant in making clear to the students that this was my own personal take on this construction. I showed them other perceptions of the project, and this became a good entry point for discussions about deep ecology and ecofeminism.

[5] I am usually distrustful of textbooks and don't like to use any textbook to drive a syllabus. It deskills the teacher and over time can create a dependence that, in its prescriptions, tends to deskill the teacher even further.

first.[6] I believe it has enough power of its own to have significant effect in even the most poorly run of classes. But it cannot function at its optimum without certain pedagogical techniques on the part of the educator: a prepared environment, a democratic classroom, and the use of critical literacy.

Prepared Environment

I believe strongly in what Maria Montessori calls "prepared environment" (Montessori 1917, xviii). The way a classroom is organized spatially provides an atmosphere for a class that helps set the tone for how that class will develop. For example, I always arrange the desks in a circle, never in rows. I have no desk of my own in the room, no delineated place from which I lecture and pass out papers. I change my seat when I can, although I find the students themselves tend to sit in the same seats over and over, and I am often forced to take the same remaining desk. I am trying to run a democratic classroom. "Liberatory pedagogy really demands that one work in the classroom, and that one work with the limits of the body, work both with and through and against those limits: teachers may insist that it doesn't matter whether you stand behind the podium or the desk, but it does" (hooks 1994, 138). What does hooks mean by "liberatory pedagogy"? In *Pedagogy of the Oppressed,* Paulo Freire (1970) sees liberatory pedagogy this way:

> A liberatory pedagogy sets itself the task of demythologizing; it regards dialogue as indispensable to the act of cognition which unveils reality; it makes students critical thinkers; it bases itself on creativity and stimulates true reflection and action upon reality, thereby responding to the vocation of men as beings. (64)

Because I am running this class as a lab science, I have divided the room into sections for lab and sections for discussion and note-taking. Once the students get accustomed to the setup, these boundaries get blurred, which I actually like. It shows they are not compartmentalizing certain things we do in the classroom and that they are comfortable enough to make small changes on their own. The most important reason for having a separate lab area is to have a place to keep ongoing experiments set up and know that they are safe from tampering.

The class must be small. I limit mine to ten students—the maximum number I can keep track of outdoors and the maximum number I have equipment for (nets, a small boat, etc.). It is also a good number for discussion—not too

---

[6] My original intent was to have the EA be the main focus of the first semester and the EE the main focus of the second semester. I had not considered how much the EA might influence the choices made for the second-semester project.

small, not too large.[7] Discussion is an important part of the class, and from the first days, the classroom environment must be set up to promote discussion. I almost never stand up in front of the group to lecture. From the beginning session, I set the tone for the class—we are in here to learn together, to explore our environment together. I will give direction to the class at times, but at no time, I tell them, do you have to agree with me. Ecology is a formal scientific discipline that sets up complex issues; it is a philosophy of life, and we construct our own reality within the context of what we will learn and discover in this class about ecology. There are assignments that will be asked of you in this class, but most are open-ended. Two of the assignments are very large: One is the writing of the EA, the other is the construction and production of an ecology experiment.

Democracy in the Classroom

The classroom can never be totally democratic. The nature of schooling is such that I, as educator, have to give grades, make certain decisions about my students, and see that they and I stick to a certain timetable. But we can consistently move toward democracy. When we do experiments, I am part of the team. When we have discussions, I am merely the facilitator and try to blend into the background as quickly as possible. When we write observations, I write also. When we need to get muddy and sweaty, I do likewise.

In a democratic classroom, everyone's presence is acknowledged and valued. When students feel that their presence is important to the conduct of the class, they begin to function collectively as well as idiosyncratically. Together with the teacher, they bring excitement to the classroom. bell hooks (1994) writes about excited classes in higher education being counter-hegemonic. When there is noise in a classroom, people moving out of their desks, voluble arguments, heated discussion, how can serious work be done? I believe that is true of high school, too, especially classes for juniors and seniors. An atmosphere of seriousness and rigor is expected and deemed essential to the learning process. On the contrary, hooks says, excitement in a classroom can "even stimulate serious intellectual and/or academic engagement" (7). When students can see the classroom as a communal place where collective effort on the part of students and teacher create and sustain a learning environment, then excitement is generated through that collective effort (8).

---

[7] The second year, and subsequent years, I was forced to take fifteen students in each class. This number proved much more difficult to manage efficiently outdoors, but I was powerless to do anything about lowering the numbers. Class size is always problematic, and the needs of administrators outweigh pedagogical ideals no matter where one teaches.

In a democratic classroom, everyone's responses and work are acknowledged and valued. For this to happen, students must learn to listen. They must learn how to hear one another. In the beginning, the teacher helps the students become aware of their own listening abilities. This doesn't mean the students have to listen uncritically or that the classroom is so open that anything can be said, but it does mean taking seriously what someone has said. It is a fundamental responsibility of the teacher to show by example how to listen to others seriously. Then when the students pick up on this example and begin to seriously hear each other, the teacher can move to "un-validate" her/himself. By that I mean, even in a democratic classroom, that the students' tendency to acknowledge another student's responses only happens when the teacher has validated that student's responses. The students—and the teacher—must learn that everyone can act responsibly in the classroom; it is not necessary for a student to have approval from the teacher every time he/she speaks in order for that speaking to be considered important.

It is vital for my students not only to understand the biological, chemical, and physical processes governing ecological systems, but to begin to grasp the complexities of the economical and political developments influencing what happens to those systems. I must make my classroom a space in which those transformative situations can take place. Maxine Greene writes, "We need spaces . . . for expressions, for freedom, . . . a public space . . . where living persons can come together, all of them granted equal worth. It must be a space of dialogue, a space where a web of relationships can be woven, and where a common world can be brought into being and continually renewed" (Greene 1984, 296). As one student said to me that first year, "I feel safe here. I feel I can say anything I want and people will listen to me." I, as the educator, must make sure all my students "feel safe." In the democratization of the classroom, each person is valued, respected, and challenged. The student learns to take risks, putting himself on the line for the rest of the group to question and critique. Learning to do this in the safe environment of the classroom prepares him for more public confrontation outside of school.

The shared aspect of *currere,* where students read each other vignettes from their work and respond to each other's writing, will not take place unless the students feel their space is protected.

Critical Literacy

Critical theory derived from the Frankfurt School, a group of philosophers and activists who worked on bringing scientific research to the study of Karl Marx's theory of social change. This school flourished in the 1930s in Germany, and in the 1940s in the United States, where several members of the school emigrated (Berger 1995, 43). The school has developed today into an attempt to bring truth and political engagement into alignment. Critical theory seeks to uncover unjust, undemocratic, unequal relations of power. These injustices may be defined and explored along social, political, and economic axes, or they may be informed by a broader provocative focus on issues of race, ethnicity, class, gender, sex, age, experience, and other factors that affect the way that power operates. Critical theory confronts a series of skeptical epistemological questions: "Do truth and goodness relate to each other and if so how? Do the fruits of knowledge embody a desire for moral action or a temptation to ethical and legal violation? If knowledge of the good does not lead to the good, what good, then, is knowledge?" (Payne 1997, 118–19). Critical reflection, a high degree of self-criticism, is imperative to critical theory. The reproduction of society, the maintenance of the status quo, is too often unquestioned and results in a conformity that allows for power structures to flourish and become stronger. A critical pedagogy invites the type of questioning suggested by Payne and other critical theorists, and looks at agency—since we understand that knowledge is constructed, we understand that those constructions can change. Therefore, how can this situation under study be changed, improved, so that the individuals victimized by this situation experience an emancipation from this particular coercion? Critical pedagogy is self-critical, political, and hopeful.

Critical thinking and writing are essential components of critical pedagogy (Morgan 1997; Giroux 1997). Critical thinking does not come easily; it demands rigorous thinking and an openness to experience. Critical thinkers must be risktakers, something that is often difficult for adolescents. To help my students develop the skills of critical thinking and writing, I worked with them on six elements of critical literacy. Whenever we would read a text, they would write something about that text, and those writings would become the jumping-off point for discussions on the text. The six elements of critical literacy I used for my ecology classes were developed in a class on cultural literacy I took with Dr. Patrick Shannon (Penn State University) in the summer of 1998. Students and professor produced this list together after many discussions. They are:

1. Does the author take up an aspect of the contemporary world?
2. Does the author place the aspect in historical context?
3. Does the author treat the aspect as a social construction?
4. Does the author appear to understand that her/his critique of this aspect of the contemporary world is incomplete?
5. Does the author hold a normative anchor?
6. Does the author suggest a project based on the critique?

I found these six points to be very helpful not only for my own writing but for my students' writing as well. It helped them realize that they did not have to agree with what the author was writing, but they must present a logical argument if they were disagreeing. Henry Giroux calls this "the discourse of textual analysis." He writes: "The political and pedagogical importance of this form of analysis is that it opens the text to deconstruction, interrogating it as a part of a wider process of cultural production; in addition, by making the text an object of intellectual inquiry, such an analysis posits the reader, not as a passive consumer, but as an active producer of meanings" (Giroux 1997, 137). I cannot control or predict how the students will listen to my discourse, how their own discourse will construct mine differently, and how other discourses—peers, family, school—will impinge on that construction. Using a list such as this I can at least help them think logically through an analysis and make judgments regarding those six questions. How their constructions play out is part of their knowledge construction. Part of the craft of teaching is learning how to dialogue with students so they have opportunities to voice their own ideologies yet can learn how to question their own constructions. I run the risk of sounding elitist if I do not—I have the right answers, I have the right critical analysis, I have a message to spread that works for everyone. This is not democratic.

Using Critical Literacy with *Silent Spring*

Can one individual make a difference? Can the personal sacrifices add up, promote systemic change? We read Rachel Carson for many reasons, but one key reason is to see that indeed one individual can make a difference, and in this case, a shy, retiring woman. Two years before she died of cancer, and just after the publication of *Silent Spring,* Carson wrote to a friend of hers:

> The beauty of the living world I was trying to save has always been uppermost in my mind—that, and anger at the senseless, brutish things that were being done. I have felt bound by a solemn obligation to do what I could—if I didn't at least try I could never be happy again in nature. But now I can believe that I have at least helped a little. It would be unrealistic to believe one book could bring a complete change. (quoted in Lear 1997, 397)

Time magazine named Carson as one of its 100 most influential thinkers of the twentieth century. In that issue, Peter Matthiessen wrote, "the damage being done by poisonous chemicals today is far worse than it was when she wrote the book. Yet one shudders to imagine how much more impoverished our habitat would be had *Silent Spring* not sounded the alarm" (Matthiessen 2000, 190).

The reading of *Silent Spring* must be contextualized. The student has to put himself in the early sixties when environmental issues were beginning to find favor with the younger generation, and the desecration of jungles with Agent Orange in Vietnam made the evening news on TV. The student also has to put himself in the place of a woman scientist trying to attack a major economic sector, the chemical production corporation. Monsanto, American Cyanamid, and DuPont all mounted vicious attacks on "a hysterical woman who had no accurate scientific data to back up her conclusions" (Lear 1997, 428–56).

This contextualizing is made somewhat easier because at the point in the class where we are reading *Silent Spring,* we are also working on Stage Three of the EA. The student is contextualizing himself in his own writing, trying to analyze why at that place at that point in time he may have acted the way he did.

One of the main purposes of the EA is to demonstrate for students how they construct their own knowledge about the environment.

> An environmentalist, a real estate developer, an artist, a hunter, and a bird-watcher are walking through wetlands. Each of them sees and responds to the wetlands in different ways. To the environmentalist, it is a source of life; to the real estate developer, it is land to be cleared so beachfront condominiums can be constructed; the artist sees it as something to paint; to the hunter, the wetlands are a cover for game; the bird-watcher sees a natural setting to explore. Thus the backgrounds and expectations of the observers shape their perceptions. (Kincheloe 1993, 108)

There is no one truth for the wetland environment. How are these perceptions of truth formed? Are these perceptions out of our control? As Kincheloe says, "Do we simply surrender our perceptions to the determinations of the environment, our context?" (109). Or can we become educated as to how our perceptions and others are formed? Using *Silent Spring,* the students and I look at Rachel Carson's perceptions of the environment and whether she is aware of these constructed "realities." Secondly, we try to analyze the perceptions of the chemical industry, the farmers, the average homeowner, and the government concerning the pesticide problems Carson delineates. Thirdly, we look at the use or nonuse of emotion in Carson's argument. Can the reader tell how Carson feels about pesticides? Does she use particular language to emphasize this? She repeats several points in the book more than once; what is her purpose in doing this? Students often are impatient with her repetitiveness, and I have even heard one student who was so frustrated with her for the repeated points that he said a male author would never have written like that. Is this a perceived flaw in her writing, or does it serve a real purpose?

Analyzing Carson's book critically allows the students to see the limitations of their interpretations. At this point in their education, they can only know so much about Carson, about her personal history and the history of the time in which she was doing her main work. It also begins to let them see that I—and they—can be comfortable with ambiguity and can actually see it as a positive force within the classroom. By example the teacher demonstrates tolerance, and the more used to critical thinking the students become, the more they absorb this tolerance themselves. Tolerance then becomes internalized, appreciated, encouraged because it allows for deeper critical thinking. The students begin to question each other's intolerances, and once again the synergy of the group becomes the correcting medium—the group instructs the group; the students become their own researchers, working on each other, analyzing each other.

These experiments with critical literacy add significantly to the students' ability to move through Stages Three and Four. Once they have attempted to analyze Carson's writing and context, they can turn those same skills on their own writing.

Sizer's School Leaving Project Idea

One technique I used the first semester came from Theodore Sizer's school-leaving exhibits in *Horace's School: Redesigning the American High School* (1992): "Be prepared to identify five birds, insects, trees, mammals, flowers, and plants from our immediate local environment" (65). I had the students find and draw these thirty organisms from their own local environments and put these drawings in their observation journals. From that point on, they were responsible for those organisms. They were to become the in-house experts on those thirty. I made a total list at the end of the assignment so that everyone could see the diversity of each student's choices and also know who would be responsible, for example, when we would encounter a monarch caterpillar on a stalk of milkweed. It was important that the organisms be drawn; in drawing, the student sees the nuances of structure and size. It wasn't important how beautiful the drawings were; what was important was their biological accuracy. We worked on this list during the month of October, the same month the students were working on their EA regressions (Stage One). I soon realized a strong connection between the two assignments. Either an organism would remind a student of some episode in his past, or an episode would provide him with content for one or more of his list of thirty species. One student brought in a gingko leaf from his front yard and started talking about *gingko biloba* and its supposed positive effects on memory. The next day another student brought in gingko biloba for the class and suggested we pass it around to help us all remember stories for our regression stages.

Flexibility

A prolonged project such as this one is unpredictable. Even if I had been doing the EA for ten years, I still could not predict how the next year would go. Students bring different contexts to the work, and if I were to become too regimented in what I expected from the EAs, I would be stifling the writer's growth process. But because schools have a built-in rigidity to them, unavoidable to some extent because of other teachers' needs and safety issues, I had to carefully orchestrate an "atmosphere of flexibility," similar to the one I try to achieve for democracy.

My original plan was to have all the students move from one stage to the next together, so they would have the advantage of sharing the same frustration and/or excitement. The discussion changed at each stage, and I felt it was helpful to the students, especially those students who were having difficulty with their first rough drafts, to hear others' comments about how they were approaching a particular stage. (We do a rough draft for each stage, then analyze that, and make changes and additions on subsequent drafts. This allows me a chance to see if the students are getting the idea of that particular stage. Some students are much more concrete in their thinking than others, and these students often have difficulty figuring out the purpose of each stage.) I was able to get the students to think together as a group as I would read excerpts from an individual student's papers or read them sections of articles on the power of place, the making of a Life List, and the use of artifacts to help promote regression. Even though I had an agenda in my head as to when I wanted each student to move from one stage to the next, having these in-class discussions, followed by one-on-one sessions with each stage, helped me remain somewhat flexible and provided the students with a perception of flexibility with their timeframes.

Billboards

Flexibility carries over to curriculum planning as well as the day-to-day pedagogy of running a classroom. The first year, I had been wrestling with the idea of how to introduce beauty as an ecological principle. As the students were working on their EAs, they and I were constantly struck by how they always used adjectives of beauty to describe places important to them. These students had daily contact with beauty because of this school's campus, but not all of them went home to beautiful neighborhoods where they could see pristine woods and encounter wildlife. I emphasized descriptive as well as scientifically accurate depictions of campus observations in our observation journals and would often read a student's paragraph with a particularly elegant description of something we saw. But I felt we were only superficially addressing the issue of beauty, instead relying on the "high" one would receive from a beautiful sight as proof that beauty was important. When I began reading *Egotopia: Narcissism and the New American Landscape* by John Miller (1997), I thought perhaps I had found a way to deepen discussions of beauty.

1. Do people have a right to beauty? If so, does *everyone* have that right?
2. Is beauty an ecological principle?
3. Is beauty always in the eye of the beholder?
4. Can beauty be legislated?
5. Looking back at your EA, are there elements of beauty in your regression and progression? How are you affected by non-beauty?

I had the students read chapter 6 of Miller's book, and we then discussed some of the little-known facts he addresses in that chapter regarding the outdoor advertising industry. Afterwards, I had the students take pictures of billboards around the proximate locale and bring them into class. We looked at where these billboards were placed, what type of product was being advertised, and whether there were any patterns we could see. As the students became aware that many of the billboards in the inner city were on residential lots, sometimes on the front lawns of houses, I revisited the question of beauty. Did the people who lived in and around these homes have a right to beauty? Why is it that you students are concerned about billboards out in the country but tolerant of billboards in the city? The discussion became frustrating, and two students finally shut it off by saying that the issue of billboards was not important in the big scheme of things, and we should just let it go.[8]

Rene Dubos once stated that our greatest disservice to our children was to give them the belief that ugliness was somehow normal. David Orr sees problems of aesthetics as signifiers of a fundamental disharmony between people and land:

> Ugliness is, I think, the surest sign of disease, or what is now being called "unsustainability." Show me the hamburger stands, neon ticky-tacky strips leading toward every city in America, and the shopping malls, and I'll show you devastated rain forests, a decaying countryside, a politically dependent population, and toxic waste dumps. It is all of a fabric. (Orr 1992, 88)

---

[8] Because the issue of billboard advertising brought up so many interesting and difficult points, I moved that section to December in following years, to allow the students to have this discussion while they were still in the process of writing their EAs. This way, I was able to help them analyze their own experiences with nature and beauty.

There were so many instances in the various EAs when students wrote of the beauty of a particular environment. I was hoping that those connections with aesthetics would help them realize the importance of beauty throughout ecosystems. Some students were able to abstract from the importance of beauty for themselves to the rights and needs of others, but many were not.

In the fourth year, a black student who lived in the inner city became so disturbed about the placing of billboards advertising liquor and cigarettes in neighborhoods, sometimes even on people's front lawns, that he used his second-semester EE work to tackle the billboard issue in his and his grandparents' neighborhoods. He wrote letters to various businesses in those areas, asking them to either take down certain billboards or to paint over advertisements that were directly painted onto their structures. One of the arguments he used in his letters was the ugliness of the boards themselves. What, he wrote, did a small child think whenever he came out of his front door and saw a large billboard on his front lawn? Whether he could read or not, the child instinctively knew that this large metal structure was part of a decaying community.

Green Buildings

Beauty is an important feature of architecture, and this holds true for green architecture as well.

We are fortunate to be located only an hour's drive from Oberlin College. The Environmental Studies department there, under the direction of David Orr, has constructed probably the "greenest" building so far built in the United States, the Lewis Center for Environmental Studies. This building was opened to the public in February 2000. The first school year of the ecology class we followed its development through newspaper articles. The second, and subsequent, school years we visited the site in early spring. It is one thing to read about green buildings and to discuss the basic criteria for deciding whether a construction is green or not; it is an entirely different thing to visit one, see it in operation, and have discussions with the people responsible for its design. Many students had written in the progression stages of their EAs about the places where they saw themselves living in the future, not only geographically but structurally as well. Several had described homes that made the best use of the environment around them. I wanted them to see *possibility*. I also wanted to revisit this idea of beauty.

Prior to the visit I spent several class periods introducing the students to the ideas behind green architecture. We concentrated on seven characteristics: an alternate energy source in the form of photovoltaic cells; an organic system that purifies and recycles wastewater on a daily basis; the use of woods from certified sustainable forests; the use of compostible fabrics; the use of non-toxic and odor-free paints and varnishes; the use of as much recycled material as possible; and natural lighting considerations.⁹ We began with a quote of David Orr's that provided us a nice segue from the study of aesthetics in billboard politics to this study of green buildings: "This building should cause no ugliness, human or ecological, someplace else or at some later time" (Orr, quoted in Ingalls 2000, B2). I asked the students if the places they saw themselves in during their progression were ugly. Of course no one did. Why not? What made your spaces beautiful? Were they designed with the environment in mind? Were they sustainable? At first, many were not sure what that meant.

Next, we looked at some case studies of actual industries that had adapted some of the green architecture characteristics. We read Michael Reis's "The Ecology of Design" (2000), David Orr's "So That All the Other Struggles May Go On" (1998), and Zoe Ingalls's "Green Building at Oberlin Is a New Dream House for Environmental Studies" (2000). The day of the visit I gave the students a worksheet to be completed at Oberlin; I asked them to focus on one of the seven aspects of green buildings and follow it throughout the building. I also asked them to write about whether they thought this particular building was beautiful: Does this building in its endeavor to treat the environment with the least intrusion possible also address the issue of beauty? Or does form follow function here?

The students, without exception, were mesmerized by the building. They couldn't ask enough questions. Because the building was still so new, many of their questions could not be fully answered. Shawn, with his great interest in hydrogen fuel cells, wanted to know all the data on the solar panels. Cody, who saw himself as a possible fashion designer, wanted to see catalogues on the different styles available in recyclable carpeting. Brendan, who was quite knowledgeable about Japanese art, couldn't get over the beauty of the panels in the auditorium made solely of pressed wheat grasses. Clint, who had done quite a bit of woodworking with his father, wanted to know what studies were done on the efficacy of office doors that had so much glass in them (to bring in more natural light). We spent over two hours having a tour and working on the assignment, left for lunch, and came back to an hour with David Orr and two fellows from the University of Texas who were visiting Dr. Orr.¹⁰ The University of Texas is presently involved in the construction of a mammoth green building over 300,000 square feet.¹¹ The discussion was so energetic and interesting we were loath to leave. We have now visited the center four times, when the building was "on line" longer, and we can see not only what

---

⁹ I chose these seven because I knew the students would see evidence of them at Oberlin.
¹⁰ One of these fellows is actually the person in charge of sustainability on the UT campus, the only university he is aware of with such a position. It is his job to make sure that everything that takes place on that campus does as little damage to the environment as possible.
¹¹ The Oberlin Environmental Studies Building is 13,600 square feet.

has been adapted or changed from the original plans, but also see what kind of impression the building makes on different groups of students.

Several students wrote that the building was not beautiful, that it was too institutional-looking. When I asked them if they were using David Orr's definition of beauty, they said they could not change their conception of beauty that easily; in order to be beautiful, one student said, something cannot be plain. (Orr writes further about his idea of beauty: "So that forces you to think about—no matter what a building looks like—if it caused ugliness where materials were taken from the earth, or if it caused people to get cancer where the materials are manufactured, or if it puts organo-chlorine compounds in the bloodstreams of the students we teach, or if its operation requires massive amounts of fossil fuels that then alter climate, you can't call the thing beautiful" [Orr in Ingalls 2000].) Ask your average high school senior if he wants to take an hour-long bus ride to see a new school building. Sure, he will be happy to have time off from school, but I guarantee he won't be too excited about where the bus is headed. Because the EA had already raised the students' curiosity about what particular spaces they could create for themselves, they saw the trip in a totally different context. Here were *possibilities*. Here were ideas commensurate with ecological principles. And these ideas were made tangible in front of them. So many of the issues we had looked at up to now seemed unfixable, close to impossible to redirect. But here were money and energy that were being spent toward sustainability in a community that welcomed it and understood it.

Zach was so taken by the experience that he decided to apply for a Strnad project[12] for his senior year in which he would look at issues of urban sprawl in Cleveland and see how green architecture might help deal with some of those issues.

The Control for My Experiment

I actually have an interesting "control" in this research. A student came into one of my ecology classes after the first semester the first year. He had not participated in the EA. He had not been a part of the slow evolution we took toward critical literacy, to certain discussion protocols, to the way we took observations once a week and wrote them up in our journals. He had been gone from the school for the entire first semester, participating in the Mountain School in Vermont, a school based on connecting the environment with education. By the time he joined my class, the class had become a synergistic group, and I actually was more concerned about Dennis being allowed

---

[12] The same type of project Adam did with his geckos. The Strnad family provides money to students each year for in-depth research projects.

to become part of this group than I was about how he would participate in the academics of the class. I talked to the students about this beforehand, stressing the importance of making the new student feel welcome. But Dennis was never to be a part of the group. He stayed on the periphery the entire semester, although he told me several times how much he enjoyed the class. He sensed he had missed out on something, he couldn't quite put his finger on what; but to me it was the process of *currere*, doing the EA.

It is also interesting to see how Dennis's time at the Mountain School, where he did not take an ecology class per se, seemed to do little, if anything, to connect him with his space. He was reluctant to share any of this with us, as was very apparent in the first weeks, when the rest of the students were so curious about his experience and he would answer their questions with the briefest responses. Perhaps there was indeed an impression made on him that will have significant impact on how he relates to the environment. That will have to wait for later study. The Mountain School is an intense experience for all our students who go there (we usually send one junior each semester), and Dennis's responses to our class may indeed be highly colored by his experience at the Mountain School. But ostensibly, at least for now, Dennis's response to the environment was probably the least intense of any of the two classes that first year. He was interested in the academics of ecology, period. Again, I feel that his nonparticipation in the EA was a large part of that behavior.

The second year, two of my students went to the Mountain School, but they did not leave for Vermont until the beginning of the second semester; they were with me for the duration of the EA. One was in each of the two classes. I wondered how that would affect them and affect the rest of their Ecology classmates. After their first month away, I heard from both of them—both were unhappy with their new science classes. The class was termed "Environmental Studies," but, at least from the reports of these two students, there was no politicization, no hands-on experimenting, and no essay writing. There were journal entries, though, and both boys, independently of each other, decided to continue in the same vein as those we had done and eventually use those pages to add to their EAs. As for the classmates they left behind, there did not seem to be much more than a hiccup notice of their absence. I found that I was missing the two boys more than the rest of the students were, mainly because both of them had made novel contributions to class discussion.

Chapter VII
What Kind of Ecologists/Scientists Are Schools Turning Out?

We cannot win this battle to save species and environments without forging an emotional bond between ourselves and nature as well—for we will not fight to save what we do not love.
  —Stephen Jay Gould, "Enchanted Evening," in *Natural History,* September, 1991

If we want to know and cannot help knowing, then let us learn as fully and accurately as we decently can. But let us at the same time abandon our superstitious beliefs about knowledge: that it is ever sufficient; that it can of itself solve problems; that it is intrinsically good; that it can be used objectively or disinterestedly.
  —Wendell Berry, *The Gift of Good Land,* 1981

This Thing Called "Ecology"

In a recent book review in the *New York Times,* the executive vice president of Fawcett Publications is quoted from a conversation he had with Dick Teresi, the book reviewer, in April 1970, right around the first celebration of Earth Day. "Y'ever heard of this ecology thing?" he asked, pronouncing the word with a long, hard "e." He admonished Teresi: "Keep your eye on this EEE-cology thing; it's going to be big!" (*New York Times Book Review,* June 25, 2000, 22). He was right, in a way. Ecology was about to become a popular word. Many schools of thought adopted it to describe anything holistically regarded (the ecology of economics, the ecology of medicine, the ecology of child psychology, for example). Ecology as a science, though, was not popular and did not become "big." To some, ecology was equated with radicalism—longhaired hippies or organic-cotton-clothed yuppies who talked of rainforest destruction and hugged redwood trees. The Unabomber, Ted Kaczynski, and his manifesto didn't help much (Chase 2000). To others, it was too "soft." What kind of science didn't have answers that were verified with equations and experiments that could easily be controlled? Schools have generally chosen one of two tacks in this eee-cology thing: either ignored it, or stuck ecology units onto biology or earth science texts. I feel that this is a mistake; ecology may be one of the most important topics students study in school. The fate of our planet depends in large part on the types of ecologists we are producing in our schools today.

In the 1860s, Ernst Haeckel, a German zoologist, was the first to use the word "ecology" to describe a science based on experimental and mathematical methods that analyzed organism-environment relations, community structure and succession, and population dynamics. Haeckel became enamored of Darwin's theory of evolution and wanted a term to describe the many-sided struggle for existence discussed in Darwin's *On the Origin of Species* (1859). The competition within species and between species created a dynamic atmosphere and varied with environment. Haeckel called this type of study "ecology." The first American textbook in ecology, written by Frederic Clements, was produced in 1905 (Kingsland in Real and Brown 1991, 1–5).

What does the study of ecology offer students? What generally is occurring in schools today in the field of ecology? This chapter attempts to answer those two questions, but, as Kingsland writes in her introduction to *Foundations of Ecology* (Real and Brown 1991), "Ecology is such a heterogeneous science that arguments about methods, approaches, and definitions of central terms are nearly impossible to avoid" (12).

Aldo Leopold says that ecology reveals how to preserve the integrity, stability, and beauty of the biotic community (Leopold 1949, 224–25). These three nouns—integrity, stability, and beauty—are social constructions and need to be understood as such. How do the students perceive beauty? Is stability measurable? What does Leopold mean by "integrity" of systems? Biologist John Janovy (1985) writes that twentieth- and twenty-first-century biologists need to ask different questions than Darwin and Seton asked. They must look at big issues: who are we, where are we going, is there an identifiable human nature?

> [T]he human species cannot live apart from the planet upon which it evolved. We share a common bond with even the most bizarre beetle of the Peruvian rain forest. A belief in that common bond might, in fact, be the most fundamental characteristic of a biologist. (Janovy, 2)

Ecologists must acknowledge that bond—the relationship between living organisms and the environment in which they find themselves, or more interestingly, choose to find themselves. That is the study of the ecologist. As an educator, I feel my role is to link "ecologist" with other human endeavors—"taxpayer," "vegetarian," "music lover." Until more of us become ecologists—many, many more of us—the fragility of the physical world will continue to increase.

How do schools educate their students to live ecologically? Are there large numbers of schools that conduct courses in ecology? If so, what is the usual sequence within those courses? Is ecology taught in the earlier grades per se, or is it adopted into Earth Days, campus clean-ups, and history lessons?

A person who earns money being an ecologist is one kind of ecologist. There are few of these fortunate individuals, although more and more corporations are seeing the need to use the special skills of the ecologist to help them deal with their long-term needs for profitability and market. If, in the teaching of ecology, I happen to foster the desire to become one these types of ecologists in a few of my students, that is fine, but I am much more concerned with everyone else—the vast majority of us who will not live ecologically unless we are educated in ecological principles and unless we have internalized these principles into our daily lives.

The seventeen- and eighteen-year-old students I am working with are beginning to understand the complexity of the world and to be less satisfied with quick fixes and black-and-white answers to complicated scenarios. Ecology is rife with dilemmas that allow them to explore the multifaceted approaches to solving these problems. In fact, it is impossible to avoid cognitive or epistemic values in ecology. The theories that develop from ecological principles are often implicitly ethical or prescriptive. Shrader-Frechette and McCoy devote a large section of their book, *Method in Ecology: Strategies for Conservation,* to the difficulties with values in ecology. Ecological theories are "implicitly prescriptive because certain normative goals are built into specifying what is 'natural' or 'healthy' for the environment"(Shrader-Frechette and McCoy 1995, 8). The inability up to now of ecological theory to predict what will happen to an environment if X happens or Y does not happen limits ecologists if they continue to think in traditional science method. They may never be able to say exactly what would happen to said environment if X happens, just as we can not exactly predict weather patterns or earthquake tremors. Chaos theory principles may bring us closer to that reality (Gleick 1987), but the interdependence of all biotic and abiotic parts of an ecosystem make for mind-boggling equations. This is part of what is missing in ecology teaching today. We cannot treat the teaching of ecology in the same way we teach physics or earth science. Einstein, Heisenberg, and the new theories of brain development and consciousness will eventually change the way physics and biology are taught. The scientific world is becoming less black-and-white. It may be ecology's role to lead the way in those less determinate ways of teaching science.

Barry Lopez (1989) writes of knowledge of environment as lacking in most Americans' lives. "Year by year, the number of people with firsthand experience in the land dwindles . . . heralding a society in which it is no longer necessary for human beings to know where they live except as those places are described and fixed by numbers" (57). Ecology teaches us that we cannot know our space strictly through data—we have to see the relationships, the interdependencies, explore them over time, know the soul as well as the physicality.

The transition to a postmodern world will entail a resurgence of moral philosophy. In order for ecological principles to be widely adopted, virtue must be refound. Our modern world has divorced us from each other, separated us both physically and virtually to the point where it is difficult for us to see connectedness to each other. If in the short term we seem to do so well in isolation with our computers, how do we bring immediacy to our arguments for ecological living? If the larger world resists this aspect of postmodernity, then the schools must be a place to start.

The Elitism of Science

Earlier on I spoke of the inbred, hegemonic tendency of schools to treat science as an elitist subject. Not everyone can "do" science, we say. You must have a certain kind of mind, let's face it—a certain level of IQ before you can possibly understand the machinations of physics, in particular. As Margaret Eisenhart says, "It is commonplace for Americans to act as if scientists are born, not made" (Eisenhart 1996, 169). Educators talk of science literacy—which is important, I don't mean to denigrate that—but when the common perception is one of scientists as pocket-protector-toting nerds who alone seem capable of understanding the theory of relativity, we science educators set major roadblocks to science education at the undergraduate and graduate levels if we don't do something to try to erase that perception. There are all kinds of scientists.

In an ethnographic study Eisenhart conducted at the undergraduate level, she looked at the teaching of physics and of environmental biology as ways of culturally producing the "way" to be a physicist or environmental biologist. She found distinct differences in the way those disciplines prepare students for positions of power and status. Physics was preparing students to be academic problem solvers headed for jobs in laboratory and university research. Environmental biology was preparing students for work as conservationists, ecologists, wildlife managers, and forest rangers.

Environmental biology, as well as ecology, environmental studies, and environmental psychology, are considered "soft" sciences, and students can be given the impression their coursework is not as rigorous as those studying the "hard" sciences of biology, chemistry, and physics. The traditional hard science approach to curriculum is a course centered on controlled, cumulative, abstract problem solving in a laboratory or classroom situation. As Eisenhart states, this is the sine qua non of "real scientists." Hard science is apolitical and absolute. Environmental biology and the other soft sciences place students in the midst of political issues by asking them to solve realistic environmental problems, often out of the school environment, in the field. The soft scientist is expected to contribute to species and habitat protection.

The curricular structure of the soft science produces "a scientist identity that is at least in some ways counter-hegemonic to the one produced in physics" (Eisenhart 1996, 175).

Eisenhart explores the use of the term "flowery bone-head," a term she finds used on some campuses in 1995 to refer derisively to a person who does not give priority to "facts" and "knowledge" defined by numbers and hypothesis testing with experimental controls, the basic hard science approach. She also finds the term used for environmentalists who seem to think more with their hearts than with their heads, who are more passionately grounded than "scientifically" grounded. The hegemony of research science promotes its own methods to some extent by spin-doctoring the contextual, socially structured elements of the environmental sciences. And certainly no eighteen-year-old male student of mine would want to be called a "flowery bone-head."

Yet here I am promoting an emotional connection with space to produce a more successful ecology course. What we as scientists have to move away from is (1) thinking all sciences are the same and should answer to the same basic goal—the production of the dispassionate laboratory scientist; and (2) thinking of science as a religion, where there is a devotion to scientific method and no place for alternatives.

John Miller, in his controversial book *Egotopia: Narcissism and the New American Landscape* (1997), writes that education has lost its way. No longer are schools preparing students to live as "public man," but instead they are:

> facilitating the private fulfillment of our individual desires. Schools, at all levels, exist to help us get better jobs, to earn greater incomes, to consume more products and services. Once, our educational institutions clearly demonstrated we were a distinct American culture defined by the concerns of public man. Today, education clearly demonstrates that we are less members of a culture than members of an economy. (37)

Miller's concept of egotopia embodies the physical changes in our environment that are largely exhibited in suburban sprawl. Our increasing narcissism and private fantasies drive this ugly manifestation of a life based on consumption.

Miller sees education as part of that consumption. Schooling becomes just another product waiting to be consumed. Schools have to compete for the time and attention of student consumers. And so educators develop pedagogies that emphasize what they think their consumers want:

> The emphasis in the schools on technology and the hardware of television monitors, satellite dishes, and computers is a smoke screen for our education system's inability

to engage, interest, and ultimately educate our students. The emphasis on gadgetry in American schools reveals their unarticulated but real decision to use television, movies, and computer games to compete with television, movies, and computer games for students' attention. (114)

The commercial, synthetic environment of postmodernism produces a world based on television and the Internet, and the world of television and computers feeds off of postmodern behaviors. At some primordial level, the time a person spends watching television, the time that person is on-line, and the time he is physically moving through the environment become melded and indistinguishable. We become an "all-encompassing modality of consumption" (Miller 1997, 117). Schools must get back—if they ever were there in the first place, Miller says—to conveying an appreciation for real scholarship. The real values of a public school education are the values of the broader society. Those values should be respect for ideas, intellectual accomplishment, self-denial, and a sense of deferment of gratification. Instead, those values today are sociability, amiable conformity, the ability to work as a team, and the life skills necessary to use a condom and balance a checkbook (115).

Miller comes down especially hard on environmentalists. Environmentalists, he says, have no vision of the future to offer the public, no exciting ideas to offer to supplant the negative doomsday messages for the future if we don't all buckle down and stop raping the Earth. He feels the message of deep ecologists and other biocentric environmentalists is ahistoric; humans can't help but impact nature, and what ecologists must do is find ways to make that impact environmentally positive:

> A future environmental vision must amount to more than planetary ecosystems functioning normally. Only an authentic and genuine vision will excite the imagination and engage the support of a restless public searching for something meaningful in which to believe. Such a vision acknowledges that environmentalism must begin at home, within the environment in which 98 percent of us spend 98 percent of our time. A future environmental vision needs to emphatically demonstrate that the manmade environment will be different and better in a world honoring ecological values. (143)

Miller's comments on the work of environmentalists and the public's perceived impression of their work brings up a struggle I face daily in the classroom: how to present ecological concepts in such a way that I get students to consider them as useful, clear, important, and personal. If I talk as Miller is writing in this doomsday approach to a failed society, students will either be turned off to such a negative picture of their world or petrified as they look to their own immediate futures and see little or no hope for redemption. If I make no attempt to comment on the media's oft-used approach to environmental activists as a group of radical hippies intent on vandalism and other violent techniques to get what they want and ignorant of economic and cultural needs, I risk turning them off completely to a philosophy they think will make them appear "odd" and non-mainstream, something most adolescents want no part of.[1] I agree with Miller that environmentalism is in dire need of innovative, creative approaches to the horrendously complex problems of overpopulation and global warming.[2] It is my hope that the EA project gets some students to begin to think politically.[3]

I do think Miller is right when he says that deep ecologists' message is ahistoric. Deep ecology is a relatively new philosophy of ecology (more on this in chapter 11) and these ecologists are working on making their message more contextual and more practical. Since human life began, humans have affected nature, and as their density increases, their tipping of the scale increases logarithmically. Deep ecologists cannot pretend human existence will ever meld into the existence of all living organisms and that humans will become inconsequential in the big picture.

Social biologist Ashley Montagu writes of the critical role education plays in enabling people to think and to take an active part in the government of their community and country. If schools abrogate their duty to teach thinking, then we put ourselves in a dangerous situation for any ideologue to take ad-

---

[1] During the first year's project, I introduced the students to Julia Butterfly Hill, the woman who lived in a redwood tree in California (Hill 2000). She was protesting the cutting of redwood trees, especially on private lands, and chose to stay in a redwood tree as a nonviolent protest of what she deemed as insufficient concern for the fate of redwoods in California. During the second year's project, I also introduced the students to Butterfly, and during the writing of their Stage Three, she came down from her tree (Hill 2000). She had received assurances from the landowners of her tree that it would not be cut. Her story is much more detailed than I am writing here, but at no point in the two years that she was in that tree did any of my students see what she was doing as noble or good or politically expedient. The "oddity" of her behavior was something my students wanted no part of.

[2] One of the reasons I believe so strongly that age eighteen is the perfect age for students to take a course in ecology is the developmental stage they are in where creativity and innovation come easily to them. It is the young student of ecology who can look at a problem with an entirely new perspective and may be one of our best hopes for new ideas that will bring about global awareness of these serious issues.

[3] The third essay we write on the McKibben article in September is from a politician's viewpoint. The politician is supposed to be in agreement with McKibben's stand on some ecological issue and is presenting possible ways to get around the problem. These essays are usually quite impractical, as the students are just beginning to look at these ecological issues in any depth and see everything as black-and-white. They also see America as being the "master of the universe" and capable of telling other countries how to deal with their environmental problems. As we move through the school year, the students generally become more practical in their thinking, and by the end of the first semester, I am beginning to hear not only more thoughtful ideas but also conversations where students are beginning to realize how complex these issues are philosophically, politically, culturally, and economically.

vantage of. He believes schools have already done that. "[I]n the Western world that ability [to think] has been strongly discouraged by the kind of 'education' to which its victims have been subjected. From first to last our schools teach their students *what* to think rather than *how* to think. The result from every point of view has been disastrous, for it leads to the death of education and inevitably to its makeshift, *instruction,* the technologization of education" (Montagu 1997, xi).[4]

In *Bitter Milk* (1988), Madeleine Grumet's excellent book on transforming curricular practice, she writes:

> Curriculum is a project of transcendence, our attempt while immersed in biology and ideology to transcend biology and ideology. Even in the most conventional science of classroom practice we can find traces of transformative consciousness, no matter how masked in apparent compliance and convention. This perception invites us to refuse to run the classroom like a conveyance, designed to transport children from the private to the public world, but to make it instead a real space in the middle, where we can all stop and rest and work to find the political and epistemological forms that will mediate the oppositions of home and workplace. (20)

It is these "oppositions of home and workplace" that I think Montagu is referring to when he says that schools neglect thinking. What kind of ecologist are we producing in schools today? In many schools, none at all.

Project Benchmark's Position on the Teaching of Ecology

In 1993, the American Association for the Advancement of Science published Benchmarks for Science Literacy: Project 2061.[5] Some one hundred fifty teachers and administrators came together to produce a document that would promote literacy in science, math, and technology. It was to be a tool for individual schools to use when planning their curricula. Recognizing their work as a process in flux—much as science, math, and technology are today—the group was careful to provide flexible guidelines. They promoted integration of the sciences, core studies showing connections among the

---

[4] Look at what is happening with the success of *Success for All* in the elementary schools, a totally prescriptive approach to the teaching of reading (American Institutes for Research 1999, 115-20). The push toward national proficiency testing also moves thinking into the background as students are more and more often required to pass multiple-choice tests in order to graduate at certain levels.

[5] Called Project 2061 because that is the next time Halley's comet will be seen on earth. The idea was that life on the planet in 2061 will be predicated on the education of the children of this generation. How could the planners not think of ecology in this context?! What the planet in 2061 will be like will depend in large part on how we care today about the ecological needs of our planet.

various sciences, math, and technology with arts, humanities, and vocational subjects.

The science of ecology was largely ignored. Oblique references were made to it:

> Understanding and appreciating the diversity of life does not come from students' knowing bits of information or classification categories about many different species, rather it comes from their ability to see in organisms the patterns of similarity and difference that permeate the living world. Through these patterns, biologists connect the multitude of individual organisms to the theories of genetics, ecology, and evolution. (American Association 1993, 101)

In the section where twelfth grade goals are delineated, the importance of diversity is discussed, though somewhat muddied:

> The variation of organisms within a species increases the likelihood that at least some members of the species will survive under changed environmental conditions, and a great diversity of species increases the chance that at least some living things will survive in the face of large changes in the environment. (105)

In the section on the interdependence of life:

> It is not difficult for students to grasp the general notion that species depend on one another and on the environment for survival. But their awareness must be supported by knowledge of the kinds of relationships that exist among organisms, the kinds of physical conditions that organisms must cope with, the kinds of environments created by the interaction of organisms with one another and their physical surroundings, and the complexity of such systems. Students should become acquainted with many different examples of ecosystems, starting with those near at hand. (115)

The twelfth grade goals for interdependence of life focus on the stability of ecosystems, the cyclic fluctuations in ecosystems, and the effect of humans on ecosystems (American Association 1993, 117).

In the section on flow of matter and energy, twelfth grade goals include the role of decomposers in ecosystems, the burning of fossil fuels' effect on the environment, and carrying capacity:

> The amount of life any environment can support is limited by the available energy, water, oxygen, and minerals, and by the ability of ecosystems to recycle the residue of dead organic materials. Human activities and technology can change the flow and reduce the fertility of the land. (121)

Out of a 418-page book, 300 of whose pages are devoted to these guidelines, these are the only allusions to ecology I could find. The same is true of most introductory biology textbooks (e.g., Miller 1991; Gottfried 1993; Campbell, Reese, and Mitchell 2000). The authors tend to put the sections on ecology at the end of the book. Gottfried's 1993 textbook, *Biology Today,* devotes pages 593–709 to "how living things interact with each other and with their environment." The very last page of text in this book is the only place where the book discusses "overpopulation and environmental problems" (708). In this last 116 pages of the book, almost half of the pages are devoted to the descriptions of different biomes, making it mainly a vocabulary section. As any teacher who has ever used a textbook knows, those last sections are rarely, if ever, used. By the time May rolls around, the teacher will find there is no way he or she can finish the entire textbook before the end of the school year.

Campbell, Reese, and Mitchell's 2000 edition of the Advanced Placement Biology textbook doesn't get around to its section on ecology until page 1024. In this new edition, the ecology section has been pushed even further toward the end, as the prior edition had considerably fewer pages. While two new sections have been added to the course, the ecology section is slotted in once again at the end.

Sylvia Mader's *Human Biology* (1998) textbook has two sections on ecology—one on ecosystems and one on population concerns—and both are placed at the end of the book (469–508). It is not until the very last paragraph of text that she discusses anything about sustainability, and she does this by talking about using draft horses to help with logging! On page 503 she lists ten ways to reduce impact on the environment, yet none of these discuss issues of culture and change in mind-set. Instead, she suggests using less laundry soap, washing clothes in cold water, and buying garden hoses made from recycled rubber. Mader's book is more often used in colleges than upper-level high school courses, making, I think, her dumbing-down of ecological concerns even more troubling.

ChemCom: Chemistry in the Community (American Chemical Society 1996) has based its course on environmental issues: water, the conservation of chemical resources, petroleum, food, nuclear products, air, and the chemical industry. It provides a fictional story of the town of Riverwood, where, throughout the book, different environmental problems beleaguer its citizens, problems that students are asked to help solve. Unfortunately, the series is published by the American Chemical Society. As a result, nowhere in the text is the chemical industry questioned. The chemical industry seems to always choose the most environmentally friendly positions, which, of course, is only true in this fictional world the ACS has created. The ACS is teaching that chemistry is part of the process of solving environmental problems (a good thing), but it treats these complex, monumental issues too simplistically. For example:

> Can we keep from doing this? Nature conserves automatically at the atomic level. Let's look at what conservation means in human terms.
>   How can we conserve our resources? That is, how can we slow down the rate at which we use them? The "three R's" of conservation are replace, reuse, and recycle. To replace a resource requires finding substitute materials with similar properties, preferably materials from renewable resources.
>   Gathering, processing, and using resources generates unwanted materials. We must also manage these materials. The waste and dispersion of nonrenewable resources may eventually pose serious threats to the well-being of our society. It is preferable that the byproducts of what we use be gathered and stored or disposed of wisely. (115)

These kinds of simplistic comments about the apparent ease with which we could solve our environmental problems ignore any cultural aspects to the problem. Nowhere are efficiency, profitability, freedom of choice, consumerism, social status, or other factors discussed as contributions to the problem. These kinds of issues do not lend themselves to technological solutions.

In short, these textbooks are ahistoric, apolitical, and silent on issues of morality and culture and tradition. When Aldo Leopold called for a "land ethic," he was saying that a person's relationship to the environment should involve moral judgments and not just economic judgments. With the individual being presented in these textbooks as a culture-free person who makes observations and utilizes the results of these so-called objective observations as the basis for rational thought, there can be no place for values and beliefs. How can you bring discussions of profit motive and personal vanity into such a text?

If the mission of schools is to prepare good citizens, how can ecology be so widely ignored in schools? A good citizen may not need to know the light and dark reactions of photosynthesis or whether a pendulum's period depends on the length of the string or the weight of the bob, but shouldn't he know what global warming is, the schools of thought about what might be causing it, and the main arguments concerning what can be done about it; how the ozone layer can be restored, what laws have been passed to help protect it, and how effective those laws have been; the crisis of freshwater availability; and how waste of all kinds is dealt with throughout the country? What about the energy crisis beginning in California in the summer of 2000 and soon to spread to other states? What are the ecological implications of these enormous demands on electrical output?

The AP Environmental Studies Test

In 1997, the Educational Testing Service (ETS) initiated a new Advanced Placement course: AP Environmental Science. The first AP test in environmental science was given in May 1998. The test was revised somewhat and

What Kind of Ecologists/Scientists Are Schools Turning Out? 125

given for the second time in May 1999. The course is designed to be the equivalent of a one-semester introductory college course in environmental science (Ertel and Scoll 1997). Because of the uniqueness of these studies as part of standard science curriculum in high schools, the course can be offered from a wide variety of departments: geology, biology, environmental science, environmental studies, chemistry, and geography. Because of this variety, emphasis areas differ. "Some courses are rigorous science courses that stress scientific principles and analysis, and that often include a laboratory component; other courses emphasize the study of environmental issues from a sociological or political perspective rather than a scientific one. The AP Environmental Science course has been developed to be most like the former" (Educational Testing Service 1999, 1).

The AP Environmental Science test is, like all AP tests, mainly multiple choice. There are one hundred multiple-choice questions and four free-response questions. According to ETS, the test is constructed so that the average test-taker would score 50 percent in each of the two sections. (ETS states that its reason for doing so is to make the test difficult enough to allow them to find "maximum information about differences in students' achievements in environmental science"—whatever that means.) Because of its large multiple-choice component, the AP course is designed around science concepts, principles, and methodologies—areas in which multiple-choice questions can be more easily formatted. The course itself is structured around the following six themes:

1. "Science is a process": Through scientific method we learn the world.
2. "Energy conversions underlie all ecological principles": The First and Second Laws of Thermodynamics drive everything in science.
3. "The Earth itself is one interconnected system": As Darwin wrote about so well over one hundred years ago, natural systems are in a constant state of flux. This, of course, is the essential root of the famous Gaia hypothesis (James Lovelock, *Gaia, a New Look at Life on Earth*, 1979).
4. "Humans alter natural systems": As Thoreau wrote about, not quite as well, over one hundred years ago, humans spend much of their lives trying to control nature (Thoreau 1846).
5. "Environmental problems have a cultural and social context": Economics drives everything, and ecology is no exception.
6. "Human survival depends on developing practices that will achieve sustainable systems": The word "sustainability" has almost become cliché in environmental studies (Educational Testing Service 1999, 2).

ETS provides a sample curriculum outline for the AP course and also a videotape of a roundtable discussion about the course, where interested people could call in and ask questions of some of the individuals who have been part of the planning committee for the course. Themes #5 and #6 are barely addressed in either the document or the video. At one point a question is raised in the video by a caller as to how much time should be allotted in the course for the political, social, and cultural issues surrounding environmental studies. The answer pretty much belies the driving philosophy of this course, or any other AP course, for that matter: There is a large body of material to cover, and you, of course, as the teacher of that course can make decisions as to how much time you want to spend on these types of issues, but it is imperative that as many of the science principles and concepts be dealt with in detail as possible. In other words, that political and social stuff gets in the way—in the way of passing the AP test, that is.

There are no environmental studies textbooks I have seen that would not agree with the six main themes of the AP Environmental Science course. They employ the same language as the Benchmarks project, and they suggest a degree of flexibility. But as anyone who has ever taught an AP course will attest, the demands to complete the suggested curriculum are so strong that it is very difficult to stray from the suggested path. (One lab in the AP Biology course has students dissecting eight different animals in one extended lab period!)

Why did ETS decide to call this course Environmental Science instead of Ecology? By definition, ecology is the study of the interrelatedness of living organisms to their environment. It is more difficult to find a widely accepted definition of Environmental Science. One textbook defines it as "primarily concerned with preventing pollution and degradation of the air, water, and soil" (Miller 1991, A7). Several other textbooks beg the question and don't define it. The word "environmentalism" has become somewhat problematic today and seems to imply the political, economic, and social issues regarding the environment, yet ETS has deliberately downplayed those aspects of ecological study. Because studying connections is "messy," ecology does not fit easily into the mold of AP curricula.

Another issue concerning the AP Environmental course and test is that very few colleges seem to be recognizing the test. A student can spend seventy dollars to take the test, score a 4 or 5 (5 being the highest), and find out that the college he or she is about to attend has no comparable course to which to transfer the credit. (Penn State and Ohio State are two good cases in point: Pennsylvania's and Ohio's largest environmental degree programs so far do not recognize the AP Environmental Science test for credit.)

Producing Ecologists

I don't believe the AP Environmental Science course produces ecologists. It produces students who have memorized a body of knowledge, spat it back for the test, and moved on. An ecologist has to take ecological principles into his life; studying ecology has to be life-altering.

Ecologist David Orr says that education can be a dangerous thing. The way we are educating students today for the most part is dangerous—we are alienating students from life by insisting on human domination and fragmentary learning as opposed to holistic learning (Orr 1992). We overemphasize success and careers. Thomas Merton once said to a group of students, "Be anything you like, be madmen, drunks, and bastards of every shape and form, but at all costs avoid one thing: success" (Merton 1985, 11). Education, he said, produces people who are literally unfit for anything "except to take part in an elaborate and completely artificial charade." When the well-educated members of government calculate GNP, do they take into account well-loved children, healthy forests, stable climates, clean water, sustainable agriculture, and biological and social diversity? Today, formal education causes students to worry about how to make a living, how to be successful before they know who they are. And most importantly to me, formal education zaps students of curiosity and wonder and passion. In fact, these qualities are discouraged, considered inferior to a cold, calm intellectualizing of the innards of an annelid or the protein packaging of viruses. Horace Mann began his campaign for public schools by promising businesses an educated labor force, individuals who would increase profit margins because they would be educated to behave and perform well in a capitalist society (Pinar et al. 1995, 610–11). We continue to do the same thing today.

Ask a student what it means to be patriotic. He will tell you all manner of things about American democracy and loyalty to the government. He will not mention anything ecological—nothing about natural resources, water, forests, wildlife. He will not mention anything about profligacy, replenishment of spent resources, or diversity issues. He will not see or even understand how his consumerism has a negative impact on the environment; it is his "patriotic duty" to spend his after-tax dollars. He sees it as patriotism to enter fully into the technology revolution without questioning where all this technology is leading us. A sound education in ecology must point up these disparities and begin the intellectual and emotional change needed to address real human needs—the need for stable families, satisfying work, loving relationships, and community.

As mentioned previously but worth repeating, Marion Namenwirth writes, "Scientists firmly believe that as long as they are not conscious of any bias or political agenda, they are neutral and objective, when in fact they are only unconscious" (quoted in A. Gough 1998, 185). The same is true for science educators.

PART TWO

A SCIENCE CURRICULUM THAT IS BOTH PHENOMENOLOGICAL AND POSTMODERN

Chapter VIII
An Ecology Curriculum That Is Both Phenomenological and Postmodern

> Phenomenology is a disciplined, rigorous effort to understand experience profoundly and authentically.
> —Pinar et al., *Understanding Curriculum*, 1995

> Postmodernism is the essential indeterminacy of human experience.
> —Patti Lather, *Getting Smart*, 1991

> The postmodern reply to the modern consists of recognizing that the past, since it cannot really be destroyed, because its destruction leads to silence, must be revisited: but with irony not innocently. . . . Irony, metalinguistic play, enunciation squared. Thus, with the modern, anyone who does not understand the game can only reject it, but with the postmodern, it is possible not to understand the game and yet to take it seriously. Which is, after all the quality [the risk] of irony. There is always someone who takes ironic discourse seriously. . . . I believe that postmodernism is not a trend to be chronologically defined, but rather an ideal category or, better still, a *kunstwollen,* a way of operating.
> —Umberto Eco, *Foucault's Pendulum,* 1989

Phenomenology in Schools

Phenomenology is concerned with essence—the essence of lived experience—and the consciousness we bring to those experiences. J. N. Mohanty calls it "a presuppositionless, descriptive science" (Mohanty 1997, 8). A phenomenologist frees himself of everyday biases and beliefs in order to observe the essence of some phenomenon and attempt to understand the meaning or essential nature of that phenomenon. A phenomenologist asks questions that focus on process rather than outcome, using process to get at the essence of situations, behaviors, experiences, programs, emotions, motivations, and learning. For example, in ecology, a phenomenologist might look at people's perceptions of ecology, at how people experience a particular ecosystem, at how individuals react to changes in environments.

A curriculum based on phenomenology focuses on *possibility* and *becoming* as goals (*currere*'s progression and analysis stages). We cannot speculate about what human beings are in themselves. Human consciousness is in constant flux. Sartre says that human consciousness can never become a substance or an objective thing. Each new experience adds to the accumulated meaning of experience for each person and sets the stage for future possibility. Phenomenological curriculum acknowledges that continuum of human development.

The EA permits the student to watch himself from a distance, to see his own evolution. He observes himself. His *possibility*, his *becoming*, fascinate him. He develops a synthesis that, for that moment, makes sense to him.

Just as the observation of an ecosystem finds its meaning in who, where, and when the observation is occurring, so does the observation of the student's life as he moves through the steps of *currere*. A phenomenological dimension of postmodernism is that events find their meaning in subjective encounters where knowledge is constructed and reconstructed with each new experience. Phenomenology seeks description of how the world is experienced by persons and what lies behind these interpretations.

When phenomenology is applied to curriculum, standardized interpretation, master narratives, and predetermined methodologies lose their applicability. Yvonna Lincoln writes:

> Phenomenologists are themselves increasingly concerned with the abstractions represented by the scientific, technological, and instrumental approaches to curriculum that prevailed during the first half of this century. The notion of the curriculum as a set of concepts, ideas, and facts to be mastered; students as empty vessels to be filled with those concepts; and pedagogy as a set of techniques to be acquired by teachers is often rejected by contemporary curricularists. (Lincoln 1992, 91)

The field of contemporary curriculum as phenomenological text realizes that there is, to use David Smith's words, "a Western cultural predisposition against intimacy with the world. The sanctity of experience which phenomenology posits runs counter to Western metaphysics" (Smith 1988, 422). It is unfortunate; phenomenology has much to offer to schooling, as it helps students understand their own and others' perceptions of knowledge. One cannot produce "knowledge of self as knower of the world" (Pinar and Grumet 1976, 38) without studying experience in a disciplined and rigorous way.

Currere is a clear example of curriculum as phenomenological text. It describes immediate, preconceptual experience (presuppositionless), and then "makes use of the phenomenological processes of 'distancing' and 'bracketing'[1] required to do so" (Pinar et al. 1995, 414). Consciousness of the lived experience is delineated and analyzed. There is a freshness to this distancing that allows the writer to see his experiences as if for the first time. Adding ecology to this process further situates the EA as phenomenological curriculum. The uncertainties and complexities of ecological issues are better addressed through phenomenology. Traditional curriculum, subject-based and

---

[1] Kincheloe (1991) describes this "bracketing" as a conscious setting aside of everyday, accepted assumptions about one's immediate apprehensions—his presuppositionless-ness.

departmentally oriented, doesn't seem flexible enough to analyze local catastrophes that can have global implications. There is no respect for borders anymore.

Postmodernism in Schools

Habermas, in his exploration of phenomenology, demonstrated that knowledge is a social construct (McLaren 1989; Mohanty 1997, 23). Poststructuralism, or postmodernism,[2] is based on this idea—we construct our own knowledge in large part due to our experiences. How are my students constructing their knowledge of the environment? How are they formulating their own ecological principles? How are they understanding the basic ecological principles I set down for them and ask them to interpret in the light of their own experiences? The complexity of ecology adds to the relativity of this knowledge. How does ecology construct and how is it constructed by poststructuralism? The poststructuralist pedagogy of ecology links education with literature and philosophy and pulls the student toward concrete experience.

At one level, Margaret LeCompte (1995) writes, postmodernism derives from the critical tradition; in rejecting old authority, the positivist canons of traditional science and epistemology, postmodernism gives voice "to voices never heard before, . . . legitimacy to ideas that had been denied, . . . [an] upending of all standards, conventions, and rules" (101). All knowledge is partial. Reality and truth are situational, negotiated, and constructed.

In critical postmodernism, Zimmerman discerns an intriguing intersection of modernity's emancipating goals, postmodern theory's decentered subject, and radical ecology's (deep ecology) vision of an increasingly nondomineering relationship between humans and nonhumans (Zimmerman 1994, 91–149). The critical postmodernist actively contests wantonly destructive social and ecological practices while simultaneously avoiding naïve primitivism and antitechnological attitudes. This gray dialectic is difficult for the students with whom I am working. They still want to see issues in black and white: Who is the good guy, the bad guy; what is the right answer; what should we as ecologists do to clean up this particular mess? I have to keep pushing them to remember the complexity of these issues by playing devil's advocate, by having them role-play different positions, by exposing them to essays by ecologists and others writing in the field showing contrarian views. There is little danger of their becoming Luddites—technology is too much a

---

[2] I am using these terms—postmodernism and poststructuralism—interchangeably. Because they are so difficult to define, there are some who would take umbrage with this. But if pressed, I am unable to differentiate between the two.

part of their lives. But I also don't want them to think of technology as "good" when they are using it in their homes and at school, and as "bad" when big corporations or governments use it to advance their profit margins and approval ratings. Instead, I want them to be able to contextualize the uses of technology, seeing it as one of several problem-solving tools. To talk to my students of a world without advanced technology is ludicrous.

A postmodern curriculum in science should be multidimensional, relational, complex, interdisciplinary, and metaphoric. When Einstein developed his theory of relativity, he set the tone for new ways of understanding the universe, ways that were no longer linear and absolute. The second law of thermodynamics became stretched. Physicist Paul Davies has written that a system that is not closed but is open to its environment can have an "exchange of matter, energy, and entropy across its boundaries where it is possible to simultaneously satisfy the insatiable desire of nature to generate more entropy and yet have an increase in complexity and organization at the same time" (Davies 1988, 10). (Ilya Prigogine [1996] comes to the same conclusion in his study of the second law of thermodynamics, or entropy.) The universe, the Gaia, is seen as a closed system, but subsystems of the universe remain open to their environments. Educator Patrick Slattery connects these issues in science to education and curriculum development: "This is a crucial element of postmodernism: radical eclecticism necessitates an openness to diverse subcultures and environments that can increase in complexity. In the same sense, the curriculum is now seen as an open system that exists in complexity"(Slattery 1995, 234). Heisenberg's uncertainty principle declared that measuring the speed of an electron changes the position of the electron; one cannot know both the speed and location of an electron at the same time. The act of measurement affects the matter being measured. The parallel in a classroom is apparent to any teacher; a classroom observer affects the dynamics of the classroom simply by her/his presence. Quantum has shown us that matter and energy are forever changing places; the world is not rigid and fixed, but is instead dynamic and in flux. The "matter" of schooling is no exception; school curricula should no longer be based on a scientific-efficiency model (Kliebard 1995, 77). Instead, science curriculum should provide an organic understanding of the universe as a single integrated system where everything is related, a kind of quantum interconnectedness. I would call this ecology.

Educators today must realize the dual role of school curriculum as regards the state of the earth: the need to inform students about the dangers of environmental pollution, a world population of 6 billion that shows no sign of declining, indiscriminate destruction of wetlands and rain forests, global warming and ozone depletion and, secondly, the opportunity to introduce students to holistic practices that "will contribute to a postmodern global consciousness that is essential for ecological sustainability" (Slattery 1995, 170). In *Curriculum for the Postmodern Era* (1995), Slattery devotes an entire chapter to issues of ecology in curriculum. Postmodern educators, he says, "understand that destruction of both the ecosphere and the human psyche are forms of violence that are interrelated" (170). The crisis lies not only with the planet but also with ourselves.

In 1992, David Orr wrote what I consider one of the most important books so far written on ecology, *Ecological Literacy: Education and the Transition to a Postmodern World*. In this book, Orr promotes a postmodern paradigm for the teaching of ecology that he dubs "ecological literacy." Orr feels that all education today must address issues of ecological sustainability. Holistic education not only focuses on the science of natural processes but also reintroduces moral philosophy throughout the curriculum and encourages experiences in education that foster virtue. He worries about the difficulties of our achieving this ecological literacy, though, because our lives are becoming more and more remote from nature. He writes:

> [T]his is the heart of the matter. To see things in their wholeness is politically threatening. To understand that our manner of living, so comfortable for some, is linked to cancer rates in migrant laborers in California, the disappearance of tropical rain forests, fifty thousand toxic dumps across the U.S.A., and the depletion of the ozone layer is to see the need for a change in our way of life. To see things whole is to see both the wounds we have inflicted on the natural world in the name of mastery and those we have inflicted on ourselves and on our children for no good reason, whatever our stated intentions. Real ecological literacy is radicalizing in that it forces us to reckon with the roots of our ailments, not just with their symptoms. For this reason, I think it leads to a revitalization and broadening of the concepts of citizenship to include membership in a planetwide community of humans and living things. (88)

Currere helped my students begin this inner questioning: How does my personal life, how do my personal choices in that life, impinge on the rest of the world? *Currere* becomes a part of Orr's postmodern paradigm of ecological literacy.

In his book *Ecology, the Ascendent Perspective* (1997), Robert Ulanowicz describes an ecology for the twenty-first century. His early chapters deal with the Newtonian-Baconian-Cartesian explanation of a deterministic, atomistic, reversible, positivistic closed universe that answers readily to knowable rules. In this type of world, A always causes B. If A does not cause B, it is because some other previously unforeseen force interfered with the system. Behavior is mechanical. We like this way of looking at the world—it allows us to understand what causes things, to say if I could only alter this one thing, then B would not happen. But, Ulanowicz says, *"life itself cannot exist in a wholly deterministic world!"* [italics in original] (145). Ulanowicz says we should be careful about the use of the word "cause." Karl Popper, in his later writings, used the word "propensities" in place of "cause." Systems have tendencies, or propensities, that have to act through "a welter of other interfering, unpredictable phenomena" (146). Ulanowicz adopts Popper's vocabulary and attempts, through stochastics, to quantify it to some extent. This is where his use of the term "ascendency" comes from (with an "e" instead of an "a"). In science, ascendency is an index that quantifies the developmental status of a living community. It combines the size or magnitude of system activity with a look at how its component processes have organized themselves (8–9). In Ulanowicz's deconstruction of the mechanistic Newtonian world, he is careful to avoid nihilism; the role of chance and indeterminancy that he insists is present in all living systems does reach a compromise between order and disorder. It is that movement toward an organization that he wants to explore using stochastics. He talks at length of Ilya Prigogine's interpretation of the second law of thermodynamics, where entropy is inherent in the large system, but small systems within the larger system can, and often do, progress toward a greater degree of organization (Prigogine 1996). He suggests that ecology is the perfect middle ground in which to explore organizational principles. "Indeed," he says, "ecology may well provide a *preferred* theater in which to search for principles that might offer very broad implications for science in general" (Ulanowicz 1997, 6).[3]

Ulanowicz sees profound difficulties arising from not viewing the natural world in this postmodern way. By clinging to a "clockwork analogy, we will preclude from our narratives all but a limited fraction of the phenomena that constitute our daily world. If Popper is correct, and forces are but a small subset of more ubiquitous entities called propensities, then we would forever remain blind to the workings of the latter if we were to remain faithful to the Newtonian-Laplacian cosmology" (Ulanowicz 1997, 149).

Students need to see that environmental problems can follow Ulanowicz's line of thinking: They are not complex, unfathomable demons best left to qualified policy makers, but instead can be studied as systems with inherent propensities. The autobiographies help simplify and localize some of these dilemmas and give the students a jumping-off point in working through ecological issues in a postmodern way.

As I read and reread through the five years' worth of EAs, I was struck with the occurrence of five common themes in the papers, themes that I would call postmodern: care in schools, insecurities and gender issues, a

---

[3] Ulanowicz, in discussing the role of phenomenology in a postmodern ecology, points up science's hold on Newtonian principles in its naming of the most postmodern of scientific theories to date: quantum mechanics. He says that quantum physics is such a radically different way of looking at nature that it is not a mechanical system at all (31–32), and should not be named "quantum mechanics."

movement away from egocentrism, politicization, and definitions of success. These themes, I feel, link easily with phenomenological and postmodern issues in curriculum. For example:

1) Has the mission of schools changed to include the teaching of morals and ethics in this relativistic world, where, if one follows the line of thinking along a continuum, even goodness is socially constructed and relativism in ethics could lead to anarchy?
2) Feminist epistemology has influenced curriculum theory and shown that insidious messages repeated often enough to students in pedagogical sites both in and out of school buildings alter their own constructions of what it means to be male or female in today's society.
3) Has the "me" generation influenced the way we develop curriculum and the way we work with students on a day-to-day basis?
4) If critical pedagogy insists on a politicization of our beliefs, how can curriculum produce political people?
5) How have definitions of success changed from Horace Mann's idea that success in life equals success in job and Foucault's concept of cultural capital and its influence on notions of success?

In Pinar's *Heightened Consciousness, Cultural Revolution, and Curriculum Theory* (1974b), James Macdonald talks of a transcendental developmental education that, he says, should answer five curricular questions: (1) how to encourage perceptual experiences; (2) how to encourage sensitizing people to others; (3) how to encourage development of close-knit community relationships; (4) how to encourage patterned meaning structures; and (5) how to encourage development of inner strength (104). The five themes I am finding in the EAs try to answer at least some of these questions—issues of caring with questions 2 and 3, insecurities and gender issues with question 2, movement away from egocentrism with questions 2, 3, and 5, politicization with question 3, and definitions of success with question 5. In that same chapter, Macdonald used ecology as an example of what he called centering, a unitary view of the world (108). Ecology offered students a chance to think holistically—a connection between an inner centering of the body and the outer concern for the fate of the planet. He added:

> It appears that any sane attempt to educate the young must deal substantively with the impact of man and technology on his own living environment, and there appears to be little hope that we can simply solve our ecological problems with the next generation of technological developments. Ecological problem solutions call for the same value search and commitment growing from the inner knowledge of what we are and what we can be. (108)

In the following five chapters, I will explore these five themes through the students' voices, both in writings and in classroom work, and through current literature on these topics. How have these issues been addressed in this ecology curriculum, especially as concerns the EA? And, more importantly, how have students perceived these issues throughout the course, particularly through the EA?

Adolescents and *Currere*

Joe Kincheloe, in his essay "Pinar's *Currere* and Identity in Hyperreality: Grounding the Post-formal Notion of Intrapersonal Intelligence" (1998) utilizes his idea of postformalism (a word that Kincheloe created in his desire to differentiate his philosophy of pedagogy from postmodern conceptions, where more emphasis is placed on relativism and less on agency and responsibility [Kincheloe 1993]) to show the revolutionary nature of Pinar's position, where who we are and what we wish to become are focal points in curriculum. As with the term "postmodernism," Kincheloe recognizes the difficulty of defining terms such as postformalism, and instead addresses his position as an "uncovering" of what it means, an elusive definition because its meaning is constantly being reinterpreted in the light of new experiences (Kincheloe 1993, 27). Because there is this flux, this idiosyncratic change for each student as he/she grows and develops, curriculum must deal with this change, moving away from a traditional, ahistoric, static posture. Kincheloe sees *currere* as a viable tool for dealing with some of the negative aspects of postmodern culture. He writes:

> A *currere* catalyzed by advances in feminist theory, post-structuralist deconstruction, and the sophistication of qualitative research strategies can provide a valuable point of departure in the study of the late twentieth century crisis of identity. Indeed, a reconstituted postmodern *currere* can help us promote a multicultural curriculum of identity that explores the genesis of our ways of seeing and the nature of our consciousness construction. (131)

He argues that *currere* can provide a "transcendence of egocentrism." With the direction of a skilled practitioner, *currere,* while helping students see themselves as they are and how they wish to become, can at the same time work to help students see the world not only in the context of themselves (Kincheloe 1993, 138). *Currere* is not meant to be solipsistic. This is a fine line for anyone to draw, and *currere* can provide a mechanism that will stretch the student to move away from himself while learning more about himself. (See chapter 11 for examples of how some of my students were able, through their use of *currere,* to do this—to move away from themselves and see themselves as part of a larger world.)

Where Pinar's first project for the use of *currere* involved adult educators examining their roles as teachers, because he was at the time working with student teachers at the University of Rochester, Kincheloe stresses the value of using *currere* with teenagers. "In an era overwhelmed by startling adolescent suicide rates (400,000 attempts per year) intrapersonal understanding takes on an even greater importance" (Kincheloe 1993, 141).

Pinar too sees the efficacy of trying *currere* with high school students and, in fact, views it as a universal method available to students of all ages. In using *currere,* schools ask the student, he says, "to interrupt his ceaseless swimming through the forms and to climb out to a high, dry place where he can pause and watch the stream flow by" (Pinar and Grumet 1976, 77).

Chapter IX
Caring in Schools

> An ethic, ecologically, is a limitation on freedom of action in the struggle for existence.
> —Aldo Leopold, *A Sand County Almanac*, 1949

When Nel Noddings wrote her widely read book on caring, *The Challenge to Care in Schools* (1992), she addressed a fundamental question for educators: How can my work, the subject I teach, my pedagogy, serve the needs of each of my students? How can I help my students to care for themselves, other humans, animals, the natural environment, the human-made environment, and the wonderful world of ideas (179)? Noddings's premise in that book was that education should be organized around centers of caring. She addressed the issue for teenagers as well as for elementary students: "For adolescents these are among the most pressing questions: Who am I? What kind of person will I be? Who will love me? How do others see me? Yet schools spend more time on the quadratic formula than on any of these existential questions" (20).

The EA gave my students the chance to explore these existential questions and work out for themselves issues of caring. In an analysis of Rachel Carson's *Silent Spring,* one of the second-year students wrote:

> Will we only do something about it [the use of pesticides] once the world is no longer fit to live in, or will we as a whole decide to end it before it gets out of control? We need to know the effect that this could have on each and every one of us now rather than later. This is why we must bring the problems to light for us all. For example, we make it mandatory for each student to learn the danger of drugs and alcohol in school. Yes, these are a problem and we should know about them and act on ending them but so is the world we live in. It, to me, is more important. We all live here and for that reason we should know how to take care of it. If only the government would take this issue for its importance and finally act on it by making the widespread knowledge a must instead of an option. If we only knew what we were doing we would then begin to realize and act on it. We all know to say no to drugs but we should all know to say no to the death of our world.

From another second-year student and an essay assignment for another section of Carson's book:

> People just don't care about the well-being of the environment. People figure as long as it doesn't affect them directly, and won't any time soon, why waste time doing something about it? People need to care more and be willing to take an initiative. Even though the problem may not affect them directly, it could affect future generations of their family if something isn't done to prevent the use of hazardous chemicals. I would take people who use these chemicals, and show them some land af-

fected by their "poison." I would also make it very clear that if something isn't done to prevent the spillage of chemicals, that the problem very likely would spread. I would encourage people with an economic stake in pesticides to try and use a more effective way of applying their product. I would also see to it that people producing these pesticides try and find a more environmentally safe product. People involved in the use of damaging chemicals are aware of the harm they cause, and know what needs to be done to preserve our world, they just don't care. People have evolved into greedy, self-centered creatures, who mostly care about their own well-being. People have no respect or appreciation for the world around, let alone a lot of people around them. . . . Until people start to care about something other than themselves, nothing can be done to help save our environment.

From an essay on a still different section of the Carson book:

The government didn't care in the later chapters because they were too busy thinking that they were doing the right thing. After the failure of the gypsy moth, they hoped to solve a fire ant problem in the southern states. They claimed that fire ants kill crops, cows, and even human beings. When in reality fire ants ate other insects, aren't they trying to get rid of those? [Vincent is referring to the ability of fire ants to be natural predators on certain insect populations.] The Department of Agriculture forgot to do their homework on this problem, all they cared about was the fact that fire ants were imported, thus "un-American" (could you blame them, it was the McCarthy Era). Once again they sprayed and sprayed with disastrous results.

Elizabeth Ellsworth has pointed out that critical theorists who want to democratize learning by providing equal access to privileged knowledge make the mistake of assuming that they know in what direction consciousness should be raised. "True dialogue is open to the possibility that people will find their own directions and reasons for choosing them" (from Noddings 1992, 32–33). I cannot dictate to my students what to care about. I can only provide a forum for them to learn about caring and value caring. The EA gives me that forum.

As we moved further into the second semester and concentrated on reading *The End of Nature* by Bill McKibben (1989), the students themselves were realizing that the EA had fundamentally challenged their views on caring. From an essay by a first-year student:

This first part about time reminded me immediately about our environmental autobiography. I thought on the way home of our puny existence on this earth. If you can conceive, which McKibben says we can't, an eternity that has already taken place before us, we can see that our actions today will not change the world to come. Now it can be a bump on the road, but will that affect anything? Julian Simon believes that same thing, but is it true? Something is bound to happen between now and in a million years. What we do now will affect the children of our children. What we should be doing instead is not dream of the future, but better lives around the world. We will not change what is ahead unless we change what is today. There are not enough people who care and what can we plausibly do? Take, for instance . . . teen smoking. That directly affects the individual, not even the community, but kids still do it. The consequences are known, but it continues to go on. The effort that groups have given, even I stress the problems to my friends who smoke, go unnoticed or disregarded. We have to first, before going on to environmental issues, go onto people issues. . . . Never in a million years or as long as we can comprehend are we going to be able to change anything unless we start from the bottom up. The bottom is the minds of people all over, enthusiasm for change and for the betterment of the universe needs to be injected in all. Race, sex, disease is not the issue unless people care. We have to make people care, but how to do that I don't know.

Rhett is not alone in that conundrum. Noddings would agree—getting students to care is difficult.

Responsibility and care come from concepts of self that are rooted in connection and relatedness to others, whether those "others" are human or nonhuman. Ecological principles only work if individuals internalize them and feel a sense of responsibility toward the environment and care about what happens to ecosystems. While these attitudes are not necessarily gendered, clearly more women than men define themselves in terms of their relations and connections to others. Along with Nell Noddings, Carol Gilligan and Nona Lyons have both traced the development of morality in women organized around notions of responsibility and care: Gilligan in her seminal work, *In a Different Voice* (1982), and Lyons, who has often worked with Gilligan, in an article published in *Harvard Educational Review* in 1983, "Two Perspectives on Self, Relationships, and Morality." *Currere* was able to bring out deep feelings about relationships in many of my male students and give them a comfortable forum in which to explore their ideas of caring. This is from a second-year student's analysis section:

All my life I've been a personal person. I love to talk about my feelings with other people, and I love finding loving relationships with people. However, I have never had that with my own family. My family and I have never been close, we never talked about our personal feelings, or gave each other hugs, or hardly ever said "goodnight." I have always had that with *other* people, with my friends and their families, and at times I have found myself wishing so deeply that I could have had that with my family from the start, and though I feel I am a usually strong person, I know it is too late for my family to transform and all of a sudden become close. My parents both grew up in Japan, whose culture does not embrace close-knit personal families. However, the problem lies in the fact that *I* was born in America, and learned about being personal here. So it is hopeless, and frankly, I would rather not change anything with my family now. I have grown accustomed over the years to our different situation, and it would be too awkward and uncomfortable to make any changes. However, when I become a father, . . . I want my children to love me dearly, and I want to say to them "I love you" and have them say "I love you" back. . . . I want to watch them grow, to observe them, to learn from them so that I may also become a better person. . . . When they are older, I want them to talk to me, to tell me about their first failed test, their first kiss, the first time they break a law, the first time they make love. And when they are grown, I want them to invite me to come visit them in their new houses, and ask me to take care of *their* children.

Byron, at age nine, etched his initials in fresh cement on a driveway in his neighborhood. The driveway was for a new neighbor, one whom Byron had not yet met. The neighbor was quite upset about the marks on his cement and tried to find out who made them. Byron writes in his regression:

> I went home that night and felt terrible. I really couldn't sleep. I thought the feeling would go away maybe after a day or two. But it didn't! Every time I went by his house my stomach dropped (and I went by his house every time I had to get to my house, or leave my house, because we live on a dead end street). So of course it was annoying, not even annoying, worse than annoying. The feeling didn't go away like I thought it was supposed to. Weeks passed and I still though of it every time I saw his house. Months passed. Almost a year passed. During that time I used to come up with reasons to tell my parents how important it was they move out of the neighborhood. These reasons were serious. Neither of my parents understood what I was going through, so they just said "no." I wanted them to understand it. I really did. I wanted them to know that I couldn't live with it anymore and I thought if we just moved out of the neighborhood then maybe I could forget about it. Maybe if I didn't have to see the house everyday I wouldn't feel so guilty. It sucked, every single time I left my house I was upset.

Byron eventually worked up the courage to tell his father. His father calmly took him to see the neighbor, Byron confessed, the neighbor treated him reasonably, and Byron was able to move on. In relating that story, Byron realized how much he had cared about that neighbor and cared about his own behavior.

Martin, a nonverbal, reserved student, added this section to his synthesis, almost parenthetically:

> I love my grandmother, who has meant so much to me. Unfortunately, she passed away this last year. For as long as I can remember my grandmother has always been there for me. When I was heavy into magic my grandmother would watch my tricks over and over when everyone else would tell me to go away. I called my grandma every night just to say hello and see how she was doing. No matter what troubles I was going through or how much stress I had during the day, when I was on the phone with my grandmother everything seemed to be all right. While I was always close to my grandma, I was the closest during the last few years of her life. Somehow she always caught the strangest diseases, yet every time, even when the doctors didn't think she would make it, she got better. I had the same feeling when she got sick the last time. However, after she had been in the hospital for a while, I got this terrible feeling about losing her and every time I visited her at the hospital I couldn't help but cry. When she did pass on, the emotions that were flying through me were like nothing I had ever experienced before. Nowadays, almost every day, I think about my grandmother and every time I feel happy and a smile comes to my face because I know she will always be with me.

Martin was realizing how much he cared for his grandmother.

Anthony looked at caring in one of his regression stories not only as human-to-human but also as human-to-environment:

> I once helped totally rebuild a front yard; helping it look like no one had done any work on it in two years, to something that should have been on the cover of "Home and Garden" magazine [sic]. I can remember the over-whelming sense of satisfaction that would take over me once a job had been completed. This feeling never occurred just once; it was many times. I began to love going to work. I did not love getting up extremely early in the morning, but I was often quietly hoping that I would get the chance to make someone's yard look just how they had envisioned. It would bring a smile to my face when I saw the face(s) of the person(s) whose yard I just contributed to refine. Even if their gratitude only consisted of extending a "thank you" to the guy who did the smallest amount of work, and just ordered all of us around, I felt content. Sometimes I would ask my fellow workers silently in my head, "Don't you care about the job you just completed? All you wanted to do was get in the truck, and head out to the next customer all for more money." I found myself believing that money is not what really counts, it is the satisfaction of knowing that I had accomplished something in its entirety, and done it successfully.

The inertia of existing social structures seems to be largely responsible for the sporadic failures in the Endangered Species Act. Whether one is a biocentric or anthropocentric ecologist (more about this in chapter 11), values of loving and caring are what ultimately will save the planet. Because love is often absent in large-scale organizations and bureaucracies, good ideas like the Endangered Species Act often get distorted and, eventually, overlooked. Perhaps before we can begin to act effectively at the grassroots level, we have to learn to think globally and love the planet.

The second-year EEs (Experiment in Ecology, the second semester's main project) in large part revolved around caring. Our campus is in the process of adding several additions to the school, and the first one began in earnest in the spring of 2000. In order to find space for this first construction (an enormous addition to the athletic wing), the architects at first wanted to project the addition on a straight line from the existing structure, which would necessitate the cutting down of a substantial amount of secondary and climax forest bordering the school lake. The first-year ecology group got involved in protesting that plan. The alternate plan was to angle the new addition more toward the student parking lot, leaving most of the lake area untouched and necessitating the removal of more primary stage woods. The second-year students were well aware of what land would be damaged by this construction, and without any real discussion about it, I noticed that quite a few of the projects chosen for the EE revolved around helping out animals who would be losing habitat from the construction. Bat houses, bird houses, wood duck nesting boxes, and deer feeders were proposed. In writings the students had to do for these projects, concern for those animals was voiced, as well as the need to encourage diversity of species on the campus. This was concrete caring. The following fall when students returned to school, I, for one, was happy to see our primitive bat houses, feeders, and nesting boxes dotting the peripheral woods, a tangible reminder that we need to look out for the welfare of all living things on our campus. This is from the regression stage of one of the nesting boxes' builders:

> Now that I think about it, I really did spend a lot of time around [the school lake] and its surroundings. I can think back to times when I was not only observing the dams created by the beavers, but the bird houses that [the school] built. These houses were intended to house the birds that were relatively native to [our area]: mallard ducks, wood ducks, and Canadian geese. It really made me happy that [the school] would have such a thing because it was neat to be able to celebrate the birds that were there. I could not really tell you in much detail at all what the houses looked like, I just know that they were there.

By the time Anthony was a senior those houses were long gone. He decided to build three. He researched what they should look like, visited the local arboretum, a huge arboretum relatively close to school that employed several bird care experiments around its ponds, talked to people there about wood duck boxes, and with the help of the woodshop at school and two friends, reintroduced bird nesting boxes at the school.

"Who am I?" "What kind of person will I be?" *Currere* allowed Anthony and many others a chance to examine those two questions. Once they started that examination, they realized how they were linked to their space, to the people around them, to a particular culture. If they didn't develop a caring attitude toward those elements, they would isolate themselves and lose connection, and it is the rare teenager who wants to live on an island.

Pinar, Reynolds, Slattery, and Taubman conclude their massive work, *Understanding Curriculum* (1995), by calling for a continued dialogue on curriculum and the need to think of curriculum as a conversation of a thousand voices, all with something to say. The authors feel these voices are united in believing that school curriculum provokes us to think critically about ourselves, our families, and society:

> The point of the school curriculum is not to succeed in making us specialists in the academic disciplines. The point of school curriculum is not to produce accomplished test-takers, so that American scores on standardized tests compare favorably to Japanese or German scores. The point of the school curriculum is not to produce efficient and docile employees for business. The point of the school curriculum is to goad us into caring for ourselves and our fellow human beings, to help us think and act with intelligence, sensitivity, and courage in both the public sphere—as citizens

aspiring to establish a democratic society—and in the private sphere as individuals committed to other individuals. (Pinar et al. 1995, 848)

Currere has allowed me to use my teaching to help create those kinds of human beings. In transcending that space between teacher and student, *currere* offers me and my students a chance to transcend ourselves.

Chapter X
Insecurities/Gender Issues

Only beings who can reflect upon the fact that they are determined are capable of freeing themselves.
—Paulo Freire, *Pedagogy of the Oppressed,* 1970

> ... Jove struck them down
> With thunderbolts, and the bulk of those huge bodies
> Made pregnant by that blood, brought forth new bodies,
> And gave them, to recall her older offspring,
> The forms of men. And this new stock was also
> Contemptuous of gods, and murder-hungry
> And violent. You would know they were sons of blood.
> —Ovid, *Metamorphoses*

The conservation movement and the study of ecology parallel the movement for women's rights. There is a movement toward liberation, a reversing of the subjugation of both women and nature. There is a common egalitarian perspective. Women work to free themselves from patriarchal constraints; environmentalists, concerned about the irreversible consequences of continuing environmental exploitation, emphasize the interconnectedness between people and nature. Both the women's movement and the ecology movement are critical of the costs of competition, aggression, and domination. There are important connections between how one treats women, minorities, and the poor on one hand, and how one treats the nonhuman natural environment on the other.

Even today in this "politically correct" world we talk of Mother Earth. She is two-faced. On one hand, she is benevolent, ripe, nurturing, peaceful, serene. She, Nature, is a kindly and motherly caring provider. Her universe is ordered and planned. On the other hand, she is wild and uncontrollable. She brings plagues, famines, terrible storms. She causes illness, destroys crops, and kills children. Disorderly and chaotic, she must be controlled.

Feminist Nancy Chodorow has an interesting take on how the term Mother Earth has permeated language surrounding the environment. Gendered identities are shaped to a large extent by different relationships with the primary caretaker, who is almost always female. This differential relationship leads men to conceive of themselves as self-contained egos with only temporary, external relation to others, while women learn to see themselves as primarily related to others. At first a boy identifies himself with his mother, but he soon learns he is sexually differentiated from her. In search of his own sexual identity, he withdraws from his mother, but experiences this process as abandonment (Freud's Oedipus complex). Angered and grieved by this perceived abandonment, he learns to fear and mistrust all women, upon whom he projects his primal relationship with his mother. Because generally he is lacking in a strong, positive relationship with his father, he sees himself as not-female. Seeking to control both the woman within (his internalized mother image) and the woman without, he blocks his feelings and defines himself as separate from others; he fears intimate relationships because the most important one in his life so far ended in pain and loss. He must gain control, then, over this unpredictable force: Woman. Because "mother" was originally identified with all of reality, he tends to regard as female the undifferentiated background against which individual entities stand out: Nature. He must gain control over her, too—Mother Nature. The domination of nature parallels the oppression of women (Chodorow 1978).

How are these male teenage students perceiving Mother Nature? Do they choose language that belies a gendered approach to nature? And do they think the fact that I am a female teacher—one of very few female teachers in this school—has anything to do with the way I conduct this class, and more particularly whether the EA could happen if the teacher were male?

In *Teaching to Transgress*, bell hooks (1994) talks about *becoming* a teacher, not teacher as a construct, something idealized and fantasized.[1] In this becoming, she works with her students to produce a class. She says that it is impossible for her to separate her body from her mind, and she repudiates the mind/body split that Foucault felt was necessary for his teaching, even though in his written work he challenged the idea of mind/body split. To hooks, she (herself) is always a presence in the classroom. She writes, "if you want to remain, you've got, in a sense, to remember yourself—because to remember yourself is to see yourself always as a body in a system that has not become accustomed to your presence or to your physicality (135). Because Foucault saw himself as a powerful white male French intellectual, he felt he could deny his body. Foucault's mind was where the power lay—not his body. But women generally are not in that position. Feminists, according to hooks, have written "about the presence of the teacher as a body in the classroom, the presence of the teacher as someone who has a total effect on the development of the student, not just an intellectual effect but an effect on how that student perceives reality beyond the classroom" (137). I, as a woman in an all-boys school, deal with that in a particularly brutal way. My body is viewed by my students and fellow male faculty as very much a part of who I am. Because I am not immediately awarded power in my position as most other male teachers in the school are, I have to pass inspection; not only

---

[1] hooks is talking about the construct of college professor, not high school teacher, but I think the parallel, while not quite as pervasive, is still there.

is my mind critiqued, but my body as well. The physicality of my presence is as much a part of who I am as is my thinking ability. It is part of my pedagogy. When I move around the room, when I sit with the students at a student desk, when I help out with lab experiments, and when I work outside alongside my students, I am seen as a working physical entity. Liberatory pedagogy demands that I work within and through the limits of my body. hooks writes, "[T]eachers may insist that it doesn't matter whether you stand behind the podium or the desk, *but it does*" (138) [my emphasis]. In my attempt not only to run a democratic classroom but to limit any uses of power, I recognize myself as a female teacher and make a conscious decision to not appear unisexual. In recognizing the limits of my own identity, I try to disrupt that objectification that is so much a part of the reproductive role of schools in maintaining status quo. Certainly my being female makes a difference to my students; whether that difference would affect how the EA played out, I am not sure. One of the students in the first-year group wrote about that in his synthesis (see excerpts from Gene's synthesis in chapter 4). He says he found it easier to write some of the things he did because I was a woman and he felt he could be more open. He is the only student, though, to write or say anything to that effect. I do think, though, that some of the discussions we had in class where students were, by that point in the process, feeling comfortable and secure in the classroom, may not have taken place if there were a male teacher in the classroom. Males at eighteen are still insecure about their place in society, and being seen as macho is important to most of them.

There are so many instances of macho behavior and language in the EAs. I will cite only a few of them here so as not to be tedious, but a reader would not have much difficulty, if any, in reading any of the EAs deciding whether the author was male or female. As these sections would appear in the respective EAs, I would either individually or in the class try to point out how gender is constructed for them. In one of Gene's regression stories, he takes time to remind me that I shouldn't worry—even after this episode, he is still "a man":

> I have had many sets of stitches and many minor injuries as a result of nature. Although I have had many unfortunate experiences in the woods [related earlier in Stage 1] I have also had a lot of experiences that I will remember for the rest of my life. Ever since I was an infant I went to Chautauqua every year. When I would take a week or two vacation there, I basically lived in the water and on the beach. I would play for countless hours in the lake with my brothers and friends. Every once in a while things would get out of hand and something would go wrong between me and my brother. Once he took off my water sock and filled it with rocks and sent it flying into the lake. As you can probably guess, I never found it, but don't worry I got him back the next day. I threw him off the dock and he broke his arm.

Nick writes in his progression, "What will I do without summer and Christmas breaks? Who will make dinner every night, wake me up if I am late or run to the store for me at the last minute? The future looks intimidating at best." I assure him he will survive—somehow.

Sam begins a lovely description of his father's tradition of making an ice rink for him and his brother every winter, but he either can't resist a chance to gloat, or he decides the paragraph is becoming too "sensitive," an ongoing worry on his part throughout the process:

> [M]y family and I constructed an ice skating rink every winter. This process may sound quite simple but when it comes down to making a good one it is very time consuming. Our first one was pretty small and not that great but as we became more experienced our rinks became larger and more elaborate. This became one of my favorite times of the year; I just loved skating around while the snow fell on my warm face. I would work all day clearing the snow and consistently spray the area with misty sheets of water. My dad would then repeat this tedious procedure through the night. I would then wake up very early the next morning to be the first one to skate on the ice. I was amazed how the procedure would work to such perfection. Later in the morning herds of people from my neighborhood would flock to my backyard to compete in vicious hockey games. I can remember one instance in which my brother cross-checked [Tom] in the throat and [Tom] couldn't talk for a few days.

He employs the same tactic several pages later in talking about forts:

> The snow was piled high but it was not sturdy enough, therefore we went to the street where large chunks of ice were. These chunks of ice provided the fort with needed support. Following this procedure we began making the tunnel to enter the fort. We did this by covering a tree with snow which made our tunnel extend further out from our fort. We then added some windows at the rear of the fort. After spraying the fort with a light mist of water we let it sit overnight.
>
> The following morning even more snow had fallen, making our fort even stronger. We gathered the new snow and constructed another tunnel to our fortress. Our fort had grown from a small pile of snow to an enormous castle. We were very pleased with our accomplishments. Therefore we celebrated by throwing numerous snowballs at cars.

Harry attempts to look at his macho behavior in several sections of his EA. He has a quick temper, and he works at analyzing why he so often reacts to situations without thinking. In this excerpt from his regression, he is just beginning the process:

> I used to show off by using my power to push this fifth grader around. He always gave in to me and took my abuse. I constantly overpowered him and threw him down into the wood chips or jumped on him. [Harry himself is in third grade at this time.] I can't remember anything about him, but this one memory sticks in my head of me creeping up on him and pushing him. He was next to the pull up bars. I pushed him from behind and his head went into the metal side poles of the pull up

bar section. It was the straw that broke the camel's back. He blew up at me crying and yelling. I purposely pushed him into that pole. I don't know why I did it, but I purposely pushed him. All of his anger towards me came out. A teacher came up to us and yelled at him for yelling at me.

Another:

It's funny the things a person remembers. I don't remember what I got that Christmas, but how I acted. I don't remember who the kid was I used to beat up, but I do remember how I treated him. It reminds me of how I acted when I went to Israel. I was seven and I was in one of the most interesting places in the world. I had traveled ten hours on a plane. We even had a tour guide the entire trip who showed us so much of Israel. The entire time I slept in the car or I ran up to military men with machine guns to get a picture of me with them. I don't remember Jerusalem or even my sister's bat mitzvah. All I remember is those guys with the guns and the tanks that I climbed into.

Anthony has many stories in his regression where he, to use his word, sounds "sensitive"—long, protracted walks through the woods, collecting leaves and wildflowers, for example. He interrupts those stories to add this:

Before I lead to my next encounter, I don't want you to think that I am a totally weird person. When I was growing up, one of my best friends was not human; he was a part of nature. He was my dog Luke. Luke was a strange breed; he was part German shepherd, and part Rottweiler. He was black and tan, and was the best dog in the whole world.

Anthony returns to his regression with several "sensitive" stories about this dog, but it was interesting to see how he felt he needed to add some macho-type relationship with a macho-type dog to assure me he was indeed male.

There are many reports in the EAs of proving one's manhood to one's peers. This is one example:

Over the stream, there were many vines. One day when we were young and on summer break, we wanted to play Tarzan. There were three vines that went over the stream and we cut them. Under the vines it was very muddy and soft, so if we fell it wouldn't hurt. (We didn't take into account if we fell into the stream bed, but we didn't.) I was the first of my friends with the guts to do it. I grabbed the vine in both hands and closed my eyes. From the first vine I made the transition to the second vine. Still with the momentum of the first time I successfully made it to the third vine and to the muddy ground. My hands hurt from the coarse vine, but I was victorious. I had successfully made it from the upper bank of one side, over the river, and to the other side that was twenty feet up. All of my friends looked up to me after that.

Sometimes the stories involved proving one's manhood to one's father:

> A tradition in my family was that every year my dad and I would play football in an event known as the "Turkey Bowl." This game involved a bunch of his friends all the way back from grade school and anyone else who had heard the word from friends. This game was a fun yet sometimes fierce battle between middle aged men wishing to relive the dreams of college days when they had their strength and athletic ability. . . . [At some point Mark talks his dad into letting him play in one of the games.] I can vaguely remember the game, but I will always remember that sight of seeing all those grown men running around. It was a special feeling for me and one which I will keep with me for the rest of my life. I felt like I was a part of something much bigger, something I had never experienced while playing with the kids on my street. And when I first caught a pass it was like I just took a small step towards becoming a man.

Another:

> As a youngster I wasn't aggressive. I didn't like to be pushed around in sports, and if I was pushed I was probably a wimp about it. My dad started this new routine with me of just getting on my nerves, calling me a wimp and whatever else he could think of including "mama's boy." It was true, I was a "mama's boy." Including the name-calling was his constant pushing me around. Not in a serious manner, but just enough for me to pop a fuse. This was the start of me becoming a man. I guess I never really realized his plan at the time. I thought he was just being a jerk. But when the next basketball season came around, I was ready. That year, I was the man. I scored all the points, I won the games, and I became the center of attention.

No matter how much we discussed the construction of male gender, I knew I was fighting a losing battle. It was one thing for me to raise awareness about the hegemonic, yet often blatant, constructions of gender in general, but I was fighting hundreds of years of ingrained patriarchy. This is from Steven's synthesis:

> My life consists of three different stages: first of all, the shy, timid individual; then the almost completely confident person; and now, the finished product. I am repelled from the shy, timid boy I once was. I don't even like to talk about this stranger. I am drawn to the self-confident person I am now. I think my brother [his identical twin] had a lot to do with my transformation. He pushed me to the limit—we were always competing. My brother has always been very confident in his abilities to perform. I think his attitude about life rubbed off on me along the way.

Yet for these many instances of machismo, there are even more instances of beautiful language to describe an environment, stories about caring and intense relationships, and stories of insecurities, sometimes conquered, sometimes not. Whether one could term these sections nongendered or "feminine," I leave for the reader to decide, but they point up the postmodern position of the multiple voices within each of us. From an analysis section:

When you are writing about the future or the past it is easy to tell a story or what you hope will happen. When you write about the present you are forced to expose your sensitive side which for many is not easy.

From another analysis:

Today, I am a nice guy but a nice guy who won't take too much crap . . . from people who aren't as nice [Here Frank asks me to help him find a better word than "crap."] My father always had stories about how he had to kick some guy out of an apartment building that he owned. He would let them stay there for months without paying rent but when it just got to be too much he didn't hesitate to kick them out. My father was tough but sensitive. That is exactly how I wanted to be and believe I am today. My father also seemed tough because he never went to the doctor. He would get sick and still continue to work. I think this is why I don't complain as much as some people do. I just put up with things the way he does.

From a synthesis:

The first pattern or transformation in my autobiography is my mind's maturity. I started out not talking about the beauty of things. For example, I can remember when I was little and my parents would say something like "isn't that beautiful," and I would respond by saying I don't know. If we watched a sunset on the beach, all I wanted to do was play in the sand or go swimming. Even in the sixth grade when my family took a trip to Disney World I was not able to see the beauty of things, therefore I hated Disney World. I wanted to go on fast rides that were exciting but these type of rides were non-existent at Disney World. I was forced to go on the rides that were slow and based on visual beauty. . . . I am now able to look at a variety of different things and enjoy them as my parents always told me I would.

I may seem like some tough jock on the outside but I am really just a nice guy just like everyone else. I believe that this is perfectly normal for a person my age. I hope that I will grow in time into a stronger, more all-around person. I have not been tested emotionally but I know that the days are coming. . . . My family and my tight friends who really know me think it is funny when I try to act tough because they know that I am just a softy. . . . My friends and my family are the things that keep me going. . . . It is very hard for me to write about these type of things but it is helping me open my mind a little bit more.

Jared, who has many humorous episodes in his EA, readily accepts the tag of "feminist" for himself and often will be the first to roll his eyes when someone makes a particularly macho remark in class. This is from his progression:

I know this is a stretch, but I was thinking, what would happen if I became president. That would truly be odd, I'm not sure how I'd do. I'd probably cause the nation to crumble and have widespread anarchy and panic, but also, if I'm president I could come up with some great new laws on the environment and on animal rights, etc. So that could be real cool. But I'm not sure Rachel would want to be the first lady. Then again, maybe she'll be the first female president, and I'd be the first husband. That'd be interesting.

There are several stories about drinking:

We did our research and knew where we could get it. [Cody is talking about alcohol here, and he is 14 years old and living in London.] It was only a forty-five minute walk, but it was going to be worth every step. We knew where we would not get caught and had a time frame for every event. Looking back on it, it is quite funny, but at the time it was no laughing matter. We were to meet at [a friend's] house at six o'clock in the evening. It was not because we particularly liked his house, or the fact that we were all close to it—I was about 20 minutes away by car. It was because he had the ever-awaited "FREE HOUSE." A blessing to all teenage boys, because for one moment in your life you were in control. Rules didn't matter and you were your own master. You were able to do anything, I mean anything. You felt like you should be complaining about your back or be warning someone that "you could poke someone's eye out with that" for you were now head of the house. [Interesting that even as Cody is about to do something very immature, he thinks about acting like his own father.] You were an adult. [Also interesting how drinking is equated with adult behavior, especially heavy drinking.] . . . truth of the matter is that you were far from an adult. In fact you couldn't be farther from it if you had tried. You would do anything and everything no matter how immature. And that is exactly what we had planned. We were going to drink for the sake of getting drunk. How truly stupid we were. We had no clue at the time, though. Oh, no, we were doing something cool. Yeah, right, "cool." More like pathetic. . . .

You know, the reason behind me telling you this is that I feel the environment is also those around you and the situations that face you everyday. These things are a constant problem. . . . When I was growing up in London I saw some things that made me open my eyes. A transition, so to speak, from innocence to adulthood. As in the famous Hemingway story "Indian camp" [sic] I was forced to open my eyes to the real world versus the sheltered and picture-perfect world painted by the people around me.

Some of the stories are quite graphic, and I felt the students were using *currere* to work through their own feelings about some situations that for them must have been very difficult and confusing. Not all the students are able to make the conclusion Cody is making in the above excerpt (although, knowing Cody, he may have been making that conclusion for my benefit).

Some students were able to talk freely about gender construction:

You know what, something that my friends have been telling me for the past few years is that I was meant to be a girl. They are only teasing, of course, but I think about those comments now and I think they say something important about me. My friends say such comments as "You're such a girl-boy," "Do you wanna go to this party and get drunk or go out with some girls and share your poetry?," "Your earrings and longer hair are very fitting to your personality." And then on the other side, I've had positive comments also. "I love you because you're in touch with your sensitive side, and that's not like most guys I know," "Yeah, you're like a girl, but that's only because you're not wild and you appreciate the company of others, so it's a good thing." I prefer to listen to those positive comments. But it's very true: I'm a girl-boy. I am not wild. I would much rather spend a quiet evening talking and chilling with a group of close friends than go to a crazy party having impersonal conversations with one group and then moving on to another group ten seconds later. . . . I am very happy with myself, and I truly believe that the combination of knowing what I am like and being happy with it will only help me in my life. I think it will take me far, it will make me succeed, and I will continue being happy to some extent.

One student, although he struggled mightily in the day-to-dayness of school with gender construction, was able to write freely about it in his EA:

> Up to that point, I had run over the thought of being homosexual many, many times in my head. (Neil was 16 at the time.) Each time the thought crossed my mind I found myself making excuses to cover it up. I was certainly accepting of Homosexuality, however, I was not ready to apply the name and context to myself. I kept telling myself that my attraction to men was just a phase, that I was young and exploring and that it would pass. After my constant cover-ups and excuses, I found myself facing up to the facts. XXXX and YYYY [two friends] had brought the subject up and now it hit me right across the face. The thought, from that point on, never left my mind. My mind must have reached the carrying capacity of stress and burden. I began sorting through it all, trying to reach a reasonable and logical answer for my attraction to men. However, I kept on returning to the conclusion that I was, in fact, gay—something I just did not want to hear. I had to begin reasoning with myself and although I was still a little skeptical of my being gay, I decided that I was going to tell someone about how I was, and had been feeling. The first person I wanted to tell was my mother. . . . It was probably the hardest thing that I have ever tried to accomplish. The more I thought about it the more stressed and upset I got and tears slowly began skating across my cheeks as I silently cried in the dark of the passenger seat. [Neil is on a car trip with his mother.] This whole thing really scared me. Up to this point, I had been able to talk to my family—immediate and extended—about anything; I knew they were accepting. Now, however, although I had such a strong bond with them, although I knew deep inside that they would accept my being gay, I was overwhelmed with fear that they wouldn't. I had never been confronted with something that I was scared and unable to talk to them about. Whereas I was usually self-sufficient and strong-headed when faced with problems such as this, I now found myself weak and afraid of the consequences. . . . Throughout the entirety of my sobbing announcement and release of pressure, my mother held my hand. She spoke in her soothing tone, trying to calm me. I still at this point did not want to be gay although I knew it was true. I kept saying repeatedly that I didn't want to be gay, that I wanted to have kids. I remember Mom asking why I didn't want to be gay. She had no problem with it, and she expressed that to me as best she could.

Kathleen Weiler, in writing about the struggle for critical pedagogy in schools, says that classrooms should be sites where consciousness and ideology can be questioned, "where critical thinking is encouraged, and where for both students and teachers, "it's okay to be human." (Weiler 1995, 122). What it means to be human may be an easier question to answer than what it means to be male or female. What makes a good man? Through the EA, my students and I were only able to begin such a dialogue. This excerpt from Brendan's regression, I believe, shows the tension in today's postmodern thoughts about maleness:

> [W]hen I was probably about ten, my uncle showed me another place that I would hold my memories in: the Japanese mountains. From our house in Hadano City, an hour outside of Tokyo, you could always look to the north and see brilliant green mountains looming on the horizon. Even in the early morning, you could still see the peaks above the mist and fog. Sometimes, on a very clear day, you could also see Mt. Fuji, the highest point in Japan. It was to this mountain that my uncle took me. We started the drive early in the morning; it was just me and my uncle, even though my mother and sisters had accompanied me on the trip. My uncle and I have always been close, even though we live halfway around the world from each other, and so making this trip alone was something I was looking forward to. I was hoping that once we got there, we would be able to walk through a particular forest, called Obake No Moli, or the Haunted Forest. Supposedly, it covered such a big area on the mountain, that many people who set their foot on it and ventured into its depth, never returned. There have been numerous accounts of ghost sightings in the forest, and because I have always been fascinated by the supernatural, I was looking for a similar experience. So it was a little bit to my disappointment that when we got to the woods my uncle just drove right past them. Still, watching the Haunted Forest flash by my eyes was enough to give me the jitters. We had driven about a quarter of the way up the mountain when my uncle veered off to the side of the road. He put the car into park and looked at me, giving me a sneaky grin. I snickered and asked him what he was doing. I looked at the road ahead of me. About thirty feet ahead was a little trail that branched off the main road. It was forbidden for cars to drive on it, but it was big enough. My uncle looked behind him to make sure no one was looking, then he slammed on the accelerator and took off on the small trail. His small Audi shook from side to side and bumped up and down on the rough trail. On either side of us were thick woods, not two feet from each side of the car. My heart was pounding and I was laughing as my uncle picked up speed. After about a minute and a half of the roller coaster ride, the trail ended and opened up into a big clearing, with gravel and bits of grass covering the ground. I understood then why cars were not allowed to drive up there. At the end of the clearing was a two hundred foot drop. As I neared the cliff, I knew that I was about to see something that would hold a permanent picture in my mind. I stepped to the edge. I looked down. I suddenly couldn't speak. For below me was about the most spectacular sight I have ever laid my eyes on. I saw the tops of woods, the top of the Haunted Forest, and I saw little streams running through the middle of the trees, and I could barely make out the little waterfalls it made, and it was this water that people drove hundreds of miles to drink, for it was considered to be the purest water in the country. And to the right, I saw the peak. The snow-covered top was as magnificent as in the pictures. My uncle, I saw, was feeling the same kind of feeling I was. He was quiet, and stared at the great expanse that lay before him. After a few minutes, we returned to the car. As we walked we were still speechless.

Insecurities/Gender Issues

Psychologists Ronald Levant and William Pollack, in their book *A New Psychology of Men* (1995), discuss the very early stages of a movement to rethink what it means to be male. A new psychology of men, they feel, will be a necessity in light of the impetus of the women's movement and the economic ramifications of a two-earner family. With the initial reading of the human genome in July 2000 and the continued development of the new field of evolutionary biology, there will come new biological knowledge, some of it relating to sex differences, that will challenge us "to integrate this new information into a more complex model of the social construction of gender" (385). There will be multiple masculinities. Pollack and Levant add:

> In this more courageous world of the new masculinities, we can expect men to become more openly connected and emotionally expressive. Consequently, men may become less anxious that exhibiting such traditional "female" traits as caring and closeness will make them "less" of a man. (387)

Currere has given my male students opportunity to examine their connection to other humans and to the environment and to be emotionally expressive in what for them today is a safe place.

Dick, a third-year ecology student, was at our school for only three years. He had transferred in his sophomore year to play football, and he was nervous about changing schools. Dick looked and walked like an athlete. He told stories in class of how he would bring up his sons (he was not going to have any daughters) and how he would work out with them every night after dinner. He had nicknames for himself that reeked of testosterone. Yet when Dick wrote his EA, he put aside his bravado and entered into the project with enthusiasm and intensity. This is from the introduction to his EA:

> I guarantee you that this is a different kind of autobiography. Over the next 29 pages, I will reflect on my past and make you feel as if you are right there beside me, becoming one with the natural world. I will take you on a roller coaster ride of emotions as one minute I seem to be strong and confident, and the next, seem insecure and helpless. In this environmental autobiography, I reveal to you my true, authentic self and I implore you to judge for yourself what kind of person you think I am. This autobiography is also different because I am not just reflecting on the past, but I was able to see myself in the future, due to the experiences I came in touch with in nature. This autobiography is heartfelt, humorous, exciting, sad, and if you become one with the text, you could find this just as fulfilling as a one-on-one encounter with nature.

Indeed.

Chapter XI
Moving Away from Egocentrism

> Consciousness itself is spurred by difference in that we gain our first awareness of who we are when we gain a cognizance of our difference from another or another's ways.
>
> —Joe Kincheloe, *Toward a Critical Politics of Teacher Thinking: Mapping the Postmodern*, 1993

Transcending the Self

At the end of the first year, I was invited to a graduation party for some of my students. Wayne's mother was there, and she corralled me into a private corner to talk about the EA. She wanted to thank me for having Wayne participate in that project. She had just read Wayne's EA a few days prior to the party and was shaken by the experience. Wayne had used his EA to explore his thoughts on more than just the environment—he has had a rough life, parents unamicably divorcing, both remarrying, Wayne forever being shuffled from one house to the other. In reading Wayne's perceptions of his young life, his mother realized how he is thinking today and why he has some of the reactions at home that he does.

Wayne was planning on showing his EA to his dad. Unfortunately, Wayne had a very shaky last half of his second semester with me, and I flunked him for the semester. He was very upset with me for not giving him a passing grade (60 percent), and it was a shame that he and I ended the school year on this note. He saw himself as working very hard that last term, and I, of course, saw just the opposite. He was constantly late to class, forever complaining about not feeling well—he thought he had mononucleosis—yet he would go out and play lacrosse every day, and play it well. He also did a poor, concrete job on papers due that term—much too concrete for that point in a student's senior year. So I knew that at the time of the party he was having mixed feelings about the EA. That made it doubly interesting to me that his mother would comment about the paper so favorably. She seemed to realize how little schoolwork Wayne had done that spring. She and I talked a lot about perception. This kind of project is a peek into the teenage mind, how different their perceptions often are, and how much they see everything as somehow related to them.

Currere allows for a transcendence of egocentrism. Joe Kincheloe writes:

The social climate created by the postmodern hyperreality with its constant commercial inducements to consume, to gratify the self, contributes to an egocentric culture. In the context of personal intelligence this cultural egocentrism holds serious consequences. Egocentrism (as opposed to connectedness) reduces our awareness of anything outside our own immediate experience . . . egocentrism tends to reduce our ability to critique the construction of our own consciousness—we cannot gain the meta-consciousness to recognize the social forces that have shaped us. . . . While post-formal teachers must make sure that students have confidence in their own perceptions and interpretations, they must concurrently work to help students overcome the tendency to see the world only in terms of self. Such self-centeredness lays the foundation for ethnocentrism, racism, homophobia and sexism. . . . We do not seek resolution, just a healthy tension between the impulses. (Kincheloe 1998, 138)

Deep Ecology

Toward the end of the school year, I introduce the students to deep ecology. Deep ecology is a branch of ecology that is biocentric—all life on earth is considered to have equal value. I have several reasons for doing this late in the year. First of all, the students are ready for it. If I had introduced it earlier in the year, they would not have had enough background information on ecology in general to understand the ramifications of a deep ecology approach. Secondly, it is a controversial philosophy. Students should know about it and begin to read some of that literature. Thirdly, and most importantly for me, because of its anti-anthropocentric approach, I hope to get the students to become less egocentric in their approaches to any problems they encounter. *Currere* allows them to begin to know themselves, to concentrate on themselves as they learn to analyze why they think the way they do, but it also shows them their position in a complex, fragile world and awakens in them through progression and analysis a desire to be a responsible member of that world. For an eighteen-year-old, though, that point needs to keep being reinforced. I hope the introduction of deep ecology will do just that.

In order to put in context where deep ecologists come from and how they have been received by mainstream ecologists, I first introduce the students to different factions within the ecology field. Ecological study today can be divided into two camps: the anthropocentric approach and the biocentric approach. Anthropocentric ecologists see humans at the center. Thoreau is a classic anthropocentrist; his writings deal with man's control and use of nature. Philosopher Teilhard de Chardin and architect Buckminster Fuller are considered to be the guiding inspirations of this branch. This approach is sometimes referred to as "shallow ecology." It is the mainstream ecological approach today, where humankind is instructed to be better stewards of the planet; for example, manufacturing plants need to produce less pollutants and to dispense the pollutants they create in a less ecologically threatening manner; or we need to produce a form of transportation that does not consume organic fuel (Sessions 1995, xii).

A biocentric approach puts the planet at the center—humans are incidental, part of the food chain. This approach is only beginning to be heard and understood by more than just the small groups who have produced this ecosophy. Their general message would be that the ecosystem is in such dire straits that we need to rethink why we have so many manufacturing plants to begin with and why we need so many cars; this approach is truly life-altering, asking humans to radically change the way they live (McKibben 1989, 175–197). There are several divisions within the biocentric camp: deep ecologists, social ecologists, and ecofeminists (Fox 1995, 269–89).

Deep ecologists, instead of looking at the industrial state and wanting to reform it as the conventional, or anthropocentric, or "shallow" ecologists want to do, ask harder questions: Where are we from? What is our relationship to the rest of the world? Are we really at the apex of evolution? Do we really need all those manufacturing plants (McKibben 1989, 181)? George Sessions and Arne Naess are probably the two most well-known deep ecologists. Earth First! is, or is not, a deep ecology organization, depending on which deep ecologists you talk to. (Deep ecologist Warwick Fox, for example, says that Dave Foreman, founder of Earth First!, has an ahistorical perspective, and his views on population control are "abhorrently simplistic" (Fox 1995, 288). Vice President Gore's attack on Earth First! as non-Christian has not been helpful to the deep ecology cause (Gore 1992, 217).

Social ecologists feel that deep ecologists are almost antihuman, and they tend themselves to be concerned primarily with issues of human social justice. Social ecologists see ecological problems as essentially political and stemming from capitalism and problems of social hierarchy and social class domination. Murray Bookchin is the leading proponent of this philosophy. His particularly combative stance has unfortunately made the social ecologist position troubling for many people (Zimmerman 1994, 151–72; Fox 1995, 276–79). (Because of the internecine warfare going on inside the social ecologist camp, I only mention this group to the students and do not concentrate on it any further.)

Ecofeminists are a much more organized camp. They reject the idea that the root of environmental crisis is anthropocentrism; instead, they point to long-standing Western cultural patriarchal attitudes of dominance over both women and Nature—an androcentrism, if you will. They see ecological problems as intraspecies rather than interspecies relationships (Warren 1997, 3). Ecofeminist Ynestra King decries deep ecologists for overlooking human suffering and taking the side of nature over culture (King 1994, 310; McKibben 1989, 181). Ecofeminists support ten "key values"—"grassroots democracy, social justice, nonviolence, decentralization, community-based economics, ecological wisdom, respect for diversity, personal and global responsibility, feminist values, and future focus" (Zimmerman 1994, 265).

I believe Aldo Leopold was dealing with deep ecology issues back in 1949 when he wrote *A Sand County Almanac* and divided land ecologists into Group As and Group Bs. The Group A conservationists saw land as soil and its function as agronomic and tree-producing. Group B conservationists saw land as living. They worried about managing natural environments for their own sake. To Leopold, they were developing an ecological conscience. Group As saw wildlife as meat and sport—how much fish caught, how many cows per acre of grazeable land. Group Bs worried about shrinking species, restoration of habitats, and the amounts of wilderness available for individuals to forage and reproduce. Leopold wrote, "In all of these cleavages, we see repeated the same basic paradoxes: man the conqueror *versus* man the biotic citizen" (Leopold 1949, 414). Clearly, the Group A member is more egocentric.

Egocentrism and Schools of Ecological Philosophy

Let us take a situation and see how deep ecologists, ecofeminists, and shallow ecologists might view it: the slaughter of whales. The International Whaling Commission (IWC) wants to halt all commercial whaling. The holistic nature of ecology would seem to call for the cessation of hunting of endangered species no matter where they are located. Recently the IWC allowed native peoples in Alaska to hunt a limited number of bowhead whales and gray whales. To further complicate this issue, scientific method shows that there is no scientific reason to ban all whaling: Most species of whale—in fact, most cetaceans—are not endangered. If the IWC bans commercial whaling, they will be doing so strictly on an ethical basis. One of the IWC commissioners stated that there was no longer any need to hunt "such large and beautiful animals for food." Deep ecologists would say whales have equal rights with people. In biocentric egalitarianism, they should not be hunted. Ecofeminists would question the power of the IWC to make regulations affecting cultures around the world. For instance, both the Japanese and the Norwegians hunt whales for food and regard whaling as an honorable profession anchored in centuries of tradition. Ignoring research that shows certain kinds of whales, such as the minke, are in no danger of extinction, the IWC, ecofeminists would say, shows an irrationality governed by racial prejudice and based on power inequities that the IWC has no interest in changing. Shallow ecologists, the mainstream ecologists, would have mixed concerns; in their desire for control over nature while making remedial changes in how certain dilemmas are mediated, shallow ecologists run the risk of becoming intractable. If an observer program is implemented, antiwhaling countries, such as the United States, will balk because they will see that putting an observer program in place will remove a barrier to whaling.

The students' first response to deep ecology is one of distrust. How could this possibly work? No American is going to be willing to give up the comforts of his life in such a radical way. At this point in the year they are still anti-Malthusian and confident in American ingenuity to solve seemingly intractable problems. However, they are not as confident as they had been in September, and some days, as the dismissal bell rings, I find individual students approaching me with questions as to just what exactly deep ecology would mean for them in their lives. They are still thinking egocentrically, but at least they are thinking, not dismissing this new idea. As for the ecofeminist position, we are no longer having discussions on any matter ecological where poverty is not addressed in some aspect. A student reporting on the Bhopal incident said that he noticed how incidents of environmental disaster invariably happen in areas of poverty. I asked him what he thought ecofeminists would do about the Bhopal incident, not only to deal with the thousands of victims (who, to this day, still have not been adequately cared for), but to prevent such incidents from recurring. It was not a fair question—ecologists in general would have difficulty answering that question. He hemmed and hawed for an answer, not yet able to understand the grave medical situation in areas such as Bhopal or the presence of large Western corporations in Third World countries.

At the same time we are discussing these divisions within the field of ecology, we are reading *The End of Nature* by Bill McKibben (1989). McKibben discusses the deep ecology camp at the end of his book. While he himself is not a deep ecologist, he wonders if maybe they might be on to something:

> It is an intensely disturbing idea that man should not be the master of all, that other suffering might be just as important. And that individual suffering—animal or human—might be less important than the suffering of species, ecosystems, the planet. It is disturbing in a way that an idea, like, say, Marxism is not. It is not all that radical to talk about who is going to own the factories, at least compared with the question of whether there are going to be factories. (McKibben 1989, 182)

He hastens, though, to add that while these biocentric approaches may seem extreme, these are extreme times. If industrial civilization is ending nature, possibly the only answer is to end industrial civilization, or at least radically transform it.

The students begin to think more deeply about deep ecology, and I ask them to imagine how America might be reconfigured if deep ecology became "popular." It is a difficult assignment, but it does move them away from their egocentric inclinations—at least for the moment. As they struggle with the writing, I tell them about recent developmental work done by the Gestalt Institute, looking at the brains of adolescents and how neural biology and learning theory can help explain some aspects of adolescent behavior. The research shows that at adolescence the brain is in a remarkably creative mode, that adolescents are less hampered by socially imposed constraints on thinking, and the chances for creative genius are much greater at this stage of development. I tell them that *they* are the hope for the world. By their exposure to the complex nature of the environment and by my sowing the seeds of biocentrism vs. anthropocentrism in their heads, it is possible that new, creative approaches to the environmental crisis will ensue. Possibly we are dealing inefficiently with this problem in Washington and in think tanks around the world. We should instead be encouraging teenagers to join the search for ways to halt the destruction.

As the school year winds down, I ask them to make any final additions/revisions to their Life List.[1] I notice that many of the revisions move the students away from thinking only about themselves.

One of the misunderstandings that has arisen about *currere* is that this emphasis on autobiographical text can become solipsistic (Pinar et al. 1995, 523). Yes, most of the writing takes place at home, where the writer in solitude thinks about himself. But when he brings these writings back to school and discusses them with his peers and knows he is sharing in the same exhausting, intense process, a connection is made between himself and the larger world. He becomes the Other and at the same time sees himself with multiple voices. Grumet calls this observing of individual identity a "chorus" (in Pinar et al. 1995, 523). David, a first-year student, was already beginning to think "deeply" when he wrote in his regression:

> One day when a couple of friends and I were walking from our cabin [David is at a camp in Canada] to the lodge where the meals were served, we ran into a black bear cub. I remember being incredibly scared, and my first instinct was to run. Luckily the cub was just as frightened as we and ran to a tree and climbed up it. I felt scared but also amazed at what I just saw. The beauty of the black bear cub was so amazing. I never had seen an animal so powerful before, even though it was a baby. My first reaction was fear and anger because the bear had come into my territory. I felt this way because I too am an animal. The cub threatened the safety of the people around me and me. My next feeling was one of sadness. I felt sad for the cub. The baby bear was searching for food with his mom and they found food in a cabin at the camp. I felt sad because I felt that the land I was on was the bears' land. Nowadays I feel that the bear is as much a part of nature as I am. If I, for example, walked on to

---

[1] At the end of the year, I give the students framed copies of their Life Lists and ask them to hang it somewhere in their dorm room next year where they can see it and remind themselves of the time when they produced this list, and of their goal to be able to check off all the items before they die.

a bear's piece of land, then I'm sure that the bear would have gotten just as defensive as I got. This experience helped me build an understanding that animals and man are going to have to live together and that this run-in with animals will happen.

Eric Mykhalovskiy writes about the concerns of using autobiographical text in sociological research. He says that he himself has been accused of being self-indulgent, narcissistic, and self-absorbed in his use of autobiography in various articles he has published. To characterize autobiography as self-indulgent makes claims about the nature of its content. "Just as the charge collapses the text's author and reader into one, so too it posits the writer's self as the text's object. This is a reductive practice which asserts that [autobiography] is about the self of the writer and no one or nothing else" (Mykhalovskiy 1997, 239). Self and society are interlocked, and, as Grumet so eloquently puts it, in chorus with each other. The EA texts are not about the students standing alone; they are about the individual in relation to others, and, in particular, in relation to various environments. At a time of greatest self-absorption, adolescents, through the power of *currere*, begin to see themselves in the context of the world and question their role in it.

Even in small, localized ways, the EA begins to move the student away from himself. A fourth-year student writes this in his synthesis:

> The main reason I have started, for the first time in my life, to do decently well in school is because this project [the EA] showed me how much I love my dad and all the great things he has given me. He has given me more than I feel I could ever give him. He has given me . . .trips around the states, life, love, dreams, and in return I have been selfish, absorbing all this and doing nothing in return. All the stories and images of what I will be in the future [progression] have relations to my dad in some way. With this project it makes what I am going to do with my life easier to realize, which is work hard, be honest with myself, and become more aware of others' needs beyond my own.

This fourth-year student could never be accused of solipsism:

> I am reminded every *moment* that so many people in this world walk around with their eyes shut. Forcing myself to retrieve memories of myself in nature from when I was young [his regression] allowed me to see the young age at which I achieved my understanding. At five years old, I saw the world in a more complex way than most my age—It's a joke among my family that I was born around age 40. I didn't become this way because I *matured* quickly, or because I conducted myself in a more *developed* manner. I became this way simply out of curiosity, asking myself the question "Why?" I knew there had to be something more out there, and through the years, I've learned and developed some kind of an understanding. . . . This project has helped me see that the past is only a memory, and the future only an idea—that a tree is not T-R-E-E, but that the tree doesn't have a name. It is more than that. *We* are the ones who attach words to nature, and *only* see them as words. We would rather eat the menu than the food. I have also learned I spent more time in nature

than I thought I had. I have used it every day in some way as a natural stress reliever after the hustle and bustle of the day is over. To listen to the sounds of nature as one would listen to a Chopin sonata, and to be able to follow the example of the tree and just simply "be," is a wonderful gift. Completing this project has inspired me to further embrace this gift and to live the future seasons of my life with care and kindness toward the earth. Mother Nature has wrapped this gift for us with the endless possibilities of the sky above us; now it is up to us to unwrap it and to thank her.

Sometimes regression stories remind students of when they were less egocentric. Bart is eight years old in this story and visiting New York City with his grandparents for the first time:

As we left the airport I strictly remember my grandmother telling us that we had to stay close to them because the city was dangerous and we needed to be protected. We arrived at a very nice hotel in the city and unpacked our bags. It was a suite at the top of the hotel and reminded me of a nice house, the ones we were used to living in back home. We unpacked and settled in. We were then to leave the comfort of the hotel and go to a "five star" restaurant. We took a cab over to the eatery that was a couple of blocks away from the hotel. We pulled up to the corner of the street and the restaurant was about 50 yards away. We all got out of the car and started walking toward the restaurant. About 10 yards from where we were dropped off, there was a man in dirty clothes who was almost trapped inside a tiny cardboard box. As I walked toward the man I saw there was a tattered old blanket, some old kitchen supplies, and a book sitting near the box. I walked up to the man and wondered what he was doing. I walked right over to the man and started to get out the phrase, "What are you looking for?" but right as I said "What" my grandmother grabbed me and said that it was not nice to bother him. As we walked to the restaurant I kept looking back at the man to hopefully see that he had found what he had lost. But he never got out of the box. At dinner I asked my grandparents what was the man looking for. My grandfather told me he hadn't lost anything but that he lived in that box. I was stunned. How could that man live in a box when we were staying at a fancy hotel and lived in a nice house in a good neighborhood, with a playground and trees and good things. I could not fathom that someone could live in an area where the conditions were so bad he had to look through garbage for food and had no bathroom and no running water. This was the first time in my life that I realized not all people were the same as me and my family and my friends.

Bart would continue to write about these awarenesses throughout his paper, one time mentioning he had forgotten about this New York story, but was very glad to have remembered it.

When Stanley watched his uncle die of cancer, a man who had taken him fishing many times, he wrote about his own struggle to sort out his sadness for himself and his empathic sadness for his uncle. This stoic boy ended up writing a poem about it, because, as he wrote, it was the only therapy he had left:

It was a brisk autumn day.
like the others—
children, rolling in leaves,
mittens and scarves.
Some mistook me for his son at the wake.
The Cancer—
like a fungus,
or a mold:
a shark,
or a rabid dog
had ravaged him from the inside.
What had started in his lungs
now ate at his brain.
Driving him, I suppose,
to call me,
early at dawn
under false names
and strange voices.
He had shown me Angling—
Slowly, then abruptly
arching the lure over my head,
and catching weeds.

He seemed to float above,
to transcend,
to have been separated
and independent
from nature itself.
When the death came,
the heart shocking loss
that was expected,
never came.
The loss more gradual,
like the slow coldness of the season,
faded into existence.
It bred itself.
Matured
from the body of teachings
left in me.

The comfort, of him,
the calm after the storm,
is not in the release—
the cleansing tears
and medicinal lamentations.
No, what was left
was pure, undiluted—
        sorrow and all.

Moving Away from Egocentrism

The tears I do not shed
are still within,
not corrosive,
but teaching.
To not drown
but comfort the soul.

Chapter XII
Politicization

> The reason why the world lacks unity and lies broken in heaps is because man is disunited with himself through failing to look at the world with new eyes.
> —Thoreau, *Walden*, 1846

A Critical Curriculum

Paulo Freire talks about the ethical duty of educators to intervene in challenging students to critically engage with their world so they can act on it (Freire 1970, 55). In this postmodern world it is not enough to study air contamination, acid rain, and carbon dioxide emission from factories and cars. We must also study how politics, economics, and culture influence how this pollution is dealt with. Questions such as who should make the rules governing the cleanup of pollution, who decides how much pollution is tolerable pollution, and who pays for bringing pollution to bearable limits need to be addressed. To sit dogmatically in the classroom and say "pollution is bad" is defeating. There is no critical engagement. Why does it seem that polluted areas generally hover in areas of poverty? Would Love Canal have happened if the area were populated with expensive homes? Why is the government today offering incredibly tempting mortgage rates to unsuspecting buyers to move back into Love Canal, a place whose name is in the process of being changed to something less threatening, when dioxin levels within the soil are still at dangerously high levels? These questions spark critical engagement.

Other questions arise from the giant oil spill at Prince William Sound. The students and I watch a video produced by Exxon Corporation on the Valdez disaster ("Scientists and the Alaska Oil Spill," 1992). I alert the students from the onset that the video comes from Exxon and that it was produced almost exactly two years after the date of the spill. In March 1999, we "celebrated" the ten-year anniversary of that event; we learned that scientists are finding damage in areas that at the time of the making of the video were considered to be fully recovered (Clark 1999). My students prove themselves eager, almost militant questioners of misinformation, rewinding the tape and pointing out Exxon's fallacies and the language used in the tape "manufactured so as to produce consent." Joyce King talks about dysconsciousness—perceptions, attitudes, and assumptions that justify inequity and exploitation by accepting the existing order of things as they appear to be. A lack of critical judgment reflects a lack of social ethics; there is a subjective identification with an ideological viewpoint that admits no fundamentally alternative vision of how things might be. There must be a constant examination of information.

The students relish finding this manipulation of information. They become questioners to such an extent that we stride the line between questioning and cynicism. I have to coach them as the year progresses in polite questioning of individuals and corporations we go out to visit. If they are to get anywhere in their quest to find out what is really going on with a particular industry, they have to learn that attack mode is never going to be effective in the long run. Just as we work together in the classroom—students as researchers, teachers and students together as researchers—so must they become co-researchers with the people we visit on-site.

I worry that this cynicism will lead to either a depression or a lethargy that is promoted by a sense of hopelessness. Freire writes of the need for a pedagogy of hope. As students learn to "read the world" and "read the word," two inseparable acts that help resist hegemony, they also need to develop a passion for their work. Freire talks about wanting to change the world not being enough—loving, clear, competent action is necessary to make a difference (Freire 1992, 17–27). Where I had difficulty with this was trying to find, in the field of ecology, "loving, clear, competent actions" that would make a difference. When issues of global warming and overpopulation are first addressed, the students react quickly with grandiose schemes to save the planet. It is only after several weeks of critical analysis that they begin to realize the complexity of these monster issues. Then the danger becomes one of shutdown. If the problems have been painted in such a way as to make them seem unsolvable, the students close down, concentrate on the writings of people such as Julian Simon (who says we are capable of correcting any problem we are confronted with because of our human ingenuity and creativity), and pretend these problems have no real relevancy for them. If the educator provides too many sides to the problem, showing how highly educated scientists can take a set of data or descriptions of an ecosystem and analyze them with very different results, the students are prone to jump to the position that makes their life the easiest. I found myself struggling with this constantly. How do I get them to see ecological concerns as relevant to their lives and as solvable problems? If I continue to emphasize the difficulty of these ecological problems, will they develop a nihilistic construction where their world is destroyed of all possibility of meaning?

Nihilism

Postmodernism itself can lead to nihilism if one is not careful. Patti Lather writes of nihilism as a "sort of 'white boy angst,' a gendered subjectivity that has deeply inscribed those parts of post-structuralism that serve as a mask with which a frustrated, defeated consciousness tries to cover up its own negativity" (Lather 1991, 115). The appeal of postmodernism for her comes in the intersection of postmodernism with the politics of emancipation—using postmodern appropriations to move beyond our own selves and to create a discourse of emancipation that can work in combination with a modernist technique such as consciousness-raising. A politics of emancipation, though, I believe, has to have a normative anchor. Is it possible that my students (for the most part) are already leading lives posited in such relativistic ways that they have no foundation, no moral anchor on which to build an emancipatory discourse?

Sociologist Max Weber moved science into the postmodern world when he wrote the following about science as vocation:

> In science, each of us knows that what he has accomplished will be antiquated in ten, twenty, fifty years. That is the fate to which science is subjected; it is the very *meaning* of scientific work. . . . Every scientific "fulfillment raises new questions"; it asks to be "surpassed" and outdated. Whoever wishes to serve science has to resign himself to this fact . . . for it is our common fate and, more, our common goal. We cannot work without hoping that others will advance further than we have. (Weber 1946 translation, 138)

Weber notes the "fragility of truth." This fragility should be cause for celebration, not for nihilistic implications. If "truth" implies a consistency, a permanency, then science must be seen as relativistic. But if postmodernism is carried to extremes, this relativism can lead to a nihilistic atmosphere where nothing has any more value than anything else and knowledge has no basis for authority. Life becomes purposeless, and there is no need to make personal commitments. I cannot abide this. There are aspects of postmodernism that I feel complement the study of ecology so well: the realization that what humans experience is a cultural construction, the acknowledgment that events and facts and individuals are not self-contained autonomous identities, the dereification of modern assumptions that allow us the possibility of constructing more viable human communities and relationships with the natural world. But there must be moral judgments and beliefs that prevent the chaos that would result if everyone believed that morality too was relative. And there must be hope; I cannot work with students in an atmosphere of gloom. I have to offer them alternatives, a way out. I cannot tell them we are doomed.

My attempts to subvert a nihilistic atmosphere in my classroom were complicated by certain biases of my own. For example, I believe we are in a perilous position regarding global warming and that we have abused the planet to the extent that it may not be able to muster enough of its own resources to counter this increase in carbon dioxide without severe restriction on further contamination. I generally try to keep my biases out of the classroom as much as I am able (a modernist position, I know), but in this case I really have to force myself to present an "unbiased" position. No matter how much I try to democratize the classroom, I am aware that I have power over my students, and the more I realize how little background they have on ecological issues, the more I realize how vulnerable they are to being taken in by whatever I have to say.

In an attempt to maintain an optimistic relativism, I try to show them how easy it is to sway opinion, yet with just grassroots movements major changes can happen even today. The students and I look at issues of progress. The hegemonic idea of progress is that progress always moves the culture forward. The outcome is important, the process only incidental. A postmodern look at progress would work in reverse. We look at dams. Dams have notoriously been considered progressive; they bring control over nature, they answer energy needs, and they provide jobs. We study the building of the Three Gorges Dam in China and contrast it with the systematic plan of old dam destruction now taking place here in the United States.[1] Where is the progress here? The Chinese would answer that despite having to move millions of people to new areas, creating new cities at the tops of agriculturally poor mountains, they are bringing the twentieth and twenty-first centuries to much of China (*Outside* magazine, October 1997). Yet here in America we are rethinking our posture on dams and beginning to tear down older dams. What about the relationships between the ecosystems and the people, animals, and plants who inhabit these areas? Can we compartmentalize this knowledge of dam construction without looking at connectedness? This is the ecological approach to dam construction/destruction.

How Does One Begin to Think Politically?

I have used a technique on this dam issue that I garnered from reading Sandra Harding (1991). Her technique points up one way to manage thinking about such complex issues. In an essay Harding wrote on gendered ways of knowing, she looks at four aspects of local resources and limitations to help show that all knowledge is partial and that ecological decisions have to be considered contextually: (1) Different cultures are located in different parts of nature; there is no such thing as homogenous nature, whether it is the Arctic, the tropics, the desert, the ocean, an earthquake-prone area, or an area susceptible to malarial epidemics; (2) Cultures have different interests in observing and explaining the same ecosystem; for example, a bog can be a site

---

[1] A plan, though, that has been put on hold since the very early days of the George W. Bush administration.

of wetland filtration, a possible area for real estate development, or a place for incubation of pests and disease; (3) Cultures can draw on different discursive traditions through which to observe and explain nature's regularities; for example, Earth can be viewed as Mother Earth, planet Earth, or seen through the lens of Christian beliefs of order and reflection of God's goodness; (4) Cultures can organize their scientific and technical knowledge in culturally different ways; such as teams working together to solve a common problem or as labs in competition with each other to be the first to cure a disease (Harding in Goldberger et al. 1996, 441–42). Whereas at first reading this may seem simplistic and commonsensical, it actually provides an excellent approach to looking at an environmental problem. We can look at the issue of building the Three Gorges Dam in China through the lens of feminist epistemology, using both global and local perspectives. What are Chinese cultural concerns about the environment? What is the relationship between industrial needs, technology, and preservation of cultures as seen by the Chinese? How does a communist government relate economic determinism with dictatorship of the proletariat in a situation such as this?

There are distinct benefits to studying ecology from a feminist perspective. Scientists generally do not question the social context of their science. Scientific method is considered sacrosanct. Scientists also are often not used to examining the assumptions we make in organizing nature. If ecologists, for example, organize nature hierarchically, what does that say for the types of questions asked in research? Is competition or mutualism stressed in ecosystems? If more research centers on the competition between species, is that so because it is reflecting patterns found in nature or because we have a socially determined view of nature? Again, if ecosystems are perceived hierarchically, then there must be domination in nature. Feminist perspective would question whether ecologists were perpetuating that domination or constructing new ways of living with nature.

Have some students moved beyond the "I know" and "I care" stages to "I want to do something about this"? Cody, a second-year senior, began writing articles on ecological issues for the school newspaper. Tom, a first-year senior and now a college senior, began working for a landscaper who uses only environmentally sound products in his work. Tom not only actively pursued this kind of employment, choosing to make less money than he could make working for a traditional landscaping firm, but he also began soliciting business for this relatively new venture, and several other ecology students have now worked for this same landscaper during the summers. Craig, the same age as Tom, was impatient with his philosophy class students during his freshman year at a California college, because they seemed to have done no personal analysis and were having difficulty with assignments where they had to examine their personal philosophies in regard to a philosopher they were studying. He decided to use the school e-mail system to demonstrate some of the work he had done on his EA and to suggest to students they begin to journalize some of their regression and progression thoughts. He reported back to me that he had received a sizeable amount of chat-room response to his work.

Rhett, from the second-year group, became very politically involved at school, and always tried to tie in some ecological issue with what he was working on. During the school year of 1999–2000, he talked to the entire student body about the mapling (tapping of sugar maple trees to make maple syrup) that is done on campus and how important it is to learn about it and be a part of the process. This was the first time in the twenty-seven-year history of mapling on this campus that a student has talked about working with the trees; every other time, the message was conveyed by the faculty member who runs the sugar house.[2]

Cameron, a second-year student, became very interested in the politics of the billboard industry. He lives in another city, and that city was revisiting its rules on the erection of billboards at the same time we were studying the problem in school. He began bringing in articles from that city's newspapers, and over a period of a couple of months, we followed their decision to gradually limit billboard construction in the city. Cameron had particularly watched a billboard near his home, an empty billboard. This huge steel construction was in front of a drugstore, bereft of any advertising for well over a year. Cameron talked to the drugstore manager about the billboard; the manager knew nothing about the status of the billboard and said he would like to see it taken down, too, if he had any say in the matter—which, of course, he didn't. As of late May (the end of school for seniors), Cameron had written the city council about the fate of that particular billboard. I suggested that he try to find out who actually put up the steel framing, and a year later the billboard framework was removed, although I can't be sure it was actually his efforts that brought the board down.[3] Cameron wrote the following in his analysis stage:

> After just finishing *Silent Spring* by Rachel Carson, I have a heightened sense of awareness about taking care of the environment. It has become an important issue for me, and although I am not really sure where I am going from here, I hope I can

---

[2] Rhett is now in a leadership position in college and has received university money to go to China this summer to study the environmental effects of the Three Gorges Dam.

[3] I had wanted Cameron to pursue this project still further by using it for his EE. I had suggested that he compose a pamphlet he could pass around his neighborhood to alert people to this billboard situation and ask for help in getting it taken down. He began working on this, doing research into grassroots organizations in Cleveland working on similar projects. After a while, though, he became discouraged, and ended up doing something totally different for his EE—helping Adam construct a school garden.

> figure it out soon. I am tired of not knowing what or where I am going, and nature helps ease the tension. The environment has never told me that it is busy, or that it does not have time for me. Whenever I am feeling down, it is always there to go to and seek peace of mind, and that is quite possibly the most important part about nature. Sadly, I do not take advantage of this opportunity nearly enough, and I often wish that I did. After all, it is not there just to look good. Nature is supposed to be enjoyed and people are supposed to have a good time in it. But, most people do not consider this a good time, and these are the people that neglect it. Again this takes me to the growing issues of protecting the environment. I wish that others felt the same way I do, because I am truly frightened about where things are going to go from here. Human beings are going to need all the help that they can get, and unfortunately, we are the only ones that can help ourselves. For this reason, the problem seems utterly hopeless and I can only pray that I am dead long before anything like the environment finally giving up the fight occurs.

Two years later, Cameron worked for the mayor's office and helped research possible alternatives for waste disposal. The city had just closed down its largest dump site and was forced to buy disposal space from surrounding communities until it could come up with a new plan. Cameron contacted me several times that summer while working on that project, and I know he will continue to keep working with that group as long as he can.

Clint, another second-year student, lived out in the country in an area that was quickly succumbing to suburban sprawl. He and his family were watching the construction of hastily conceived houses all around them. Clint and his father decided to begin attending township meetings and voicing their concern about the builders' supposed unconcern about environmental damage. As of the end of the school year, they had attended two meetings and spoken at both.

At the end of this five-year experiment with ecology, I was able only in a very limited way to get some of my students doing anything political. The correspondence and visits I have had from past students have been encouraging, and every student reports how much he misses the ecology class, the synergy of the group, and the opportunity for meaningful discussion. Several students still send me via cyberspace articles from on-line articles, so I know they are keeping up to some extent with what is happening environmentally. But I cannot say I have been successful in politicizing some of these students or indeed that the class as a whole has been successful. Carson, a first-year student, wrote this in his synthesis:

> I do not see myself getting into the politics of environmental issues, although I did do my sophomore English term paper on the Salvage Logging Bill, its devastating effects, and how it was being misused. I actually found this topic to be very interesting. It dealt with the fact that in America, as in most developed areas, we spend so much time preventing and extinguishing forest fires. Fires, however, are one of the healthiest and most replenishing events that can happen to a forest if they are left to burn on the natural cycle. When we prevent these fires, and dead brush builds up,

when they finally do burn unexpectedly the effects destroy the majority of the forest rather than help it. Well, the U.S. government finally discovered this (or at least acknowledged it as a problem) and they proposed the salvage logging bill.. . . . Anyway, it was a big mess, and I thought it was very interesting. This shows that I really have no idea what might pop up in my life politically that may capture my interest. This obviously did, but I foresee nothing political in my future.

Vincent, a second-year student, had political ideas of his own in his progression:

Since my days in Hawaii, I wanted to be an environmental CEO. I credit my teachers for my desire to help the Earth. [Vincent went to a special school for dyslexics in Hawaii and had two women teachers he dearly loved—one in particular was very environmentally conscious.] To help the environment as much as I can and make a buck at the same time. I don't like how MNCs (Multi-National Corporations) run things currently. [Vincent was taking AP Economics concurrently with Ecology, and he knew—and loved—all the economics lingo.] They will skip environmental laws by moving facilities to another country. I plan to be different, don't get me wrong I don't plan to be a saint. Many facilities will be in third world countries to save labor costs, but I do plan to make them environmentally friendly plants. In addition, I plan to pay more than the average corporation, but it will still be relatively cheaper than labor costs here. Basically, it is the law of supply and demand: there is just more supply there, but I am a bit more humane than MNCs. Besides, if sudden interest emerged about cheap labor, my company's PR would suffer.

Tom saw his progression through a different lens:

Other times, I dream of breaking conformity—living by my own standards and being radical. I want to be like the black civil rights activists in the early 60s, who stood up and fought for what in their hearts they knew was right. I don't know what it is, but I feel the need to sacrifice myself, lay my body on the line for a worthy cause.

A particularly introspective fourth-year student wrote in his synthesis:

Throughout my adventures in the past, the adventures I am taking now, and the ones to come in the future, I feel an obligation to make the environment very personal. Looking back, I can see that my childhood interests gradually grew with time to become an overpowering force that I will accept with open arms. I do want to learn via experience. I do not discourage studying a distant culture or environment through a book, but I will not accept someone telling me they are a master in that study if they have no hands-on experience. Books and school do a certain justice but they do not supplement being immersed in a culture or environment.

I do not know how much I can do for the environment, but I know that I will try in whatever way I can. The first step is to live a sustainable life with little impact on the earth. I hope to build a "green" house—not the plant house, but one that is completely or nearly self-sufficient in areas like electricity, sewage, water, and land degradation. . . . I have found that Germany is one of the leading countries in this development, which makes it a leading candidate for my "things to do" list. I hope to live in Germany so I can study their building design theories and learn a little about the cultures some of my ancestors came from. My continual disagreement with this country increases daily, and my expatriate feelings escalate. I need to get away from American culture where the athlete is cherished and the teacher is unappreciated, where the car you drive is the person you are, and where the petty things you own are of greater importance than the land you inhabit.

Ecologist David Orr wonders, too, if education does not get in the way of living by ecological principles:

> There is a shelf of dust-laden studies about the difference education makes. The conclusions, given present tuition rates, are remarkably ambivalent. For the majority, peer influences seem to be a more important source of ideas and behavior than professors or courses. Most students seem to regard education as a ticket to a high-paying job, not as a path to a richer interior life, let alone one of saving the planet. (Orr 1992, 150)

Other pedagogical sites, he argues, do more to educate us about living ecologically than schools do: television, automobiles, cheap energy, and so on. If schools are to make a difference in how we view sustainability and humanity and the future of the planet, we must reconceptualize. As Orr says, "This will require a serious effort to rethink the substance and process of education" (Orr 1992, 152). *Currere* can be a major step in that process.

Chapter XIII
Definitions of Success

> [In *currere*] students explore their place in the social hierarchy of their peer groups, their romantic relationships, their vocational aspirations, their relationships with teachers, and their definition of success.
> —Joe Kincheloe, in Pinar's *Curriculum: Toward New Identities,* 1998

Every fall we have a parent night at school where the parents spend the evening in the classroom, going through their sons' schedules. Teachers use this time in myriad ways, but mainly to acquaint the parents with the goals and expectations of the different courses. In autumn 1999, I ran into a parent friend of mine as parent night was ending. He was quite frustrated. Every teacher he heard that evening, he said, talked about assessment, how his particular course helped the students get better scores on the AP (Advanced Placement) tests, and how important his course looked on transcripts for college application. No one was talking about making his two sons better citizens or better people. He said what frustrated him most was that all the parents sat quietly through these ten-minute lectures and never once questioned anything. They seemed to tacitly accept these institutional assumptions, possibly because these assumptions were being defined by the school professionals themselves. Was past practice cloaking itself as theory and legitimizing these teachers' work? I thought of Foucault's idea of language manufactured so as to produce consent (Foucault 1980) and I was encouraged that at least someone was thinking critically. What does this use of language say about what makes for a successful school, a successful student? How do we educators define success today? How do parents define success for their children? In the EA, many students defined success for themselves—at least as they were understanding it at eighteen years of age.

When even the Webster *New World College Dictionary* defines success as "the gaining of wealth, fame, rank, etc.," (3rd edition, p. 1337), one can understand the frustration of a parent who obviously has different views about what success means. I wish this parent could have read some of the EAs. After the student has gone through the progression stage and begins to work on his analysis, he inevitably has to think about whether the future he saw for himself was one he was excited about, and what success meant for him. The Webster's definition was no doubt appealing to many, but they often saw the crassness of such a definition, too. Following are several examples of the students' definitions of success:

> For anyone to say they are completely happy sounds outrageous. There are things that I wish could be different, but I love my life. I've always reminded my mother a

lot of her sister, and she constantly worries if I'm depressed. I would never do what my aunt did [she committed suicide at quite a young age]. I understand she felt there was no escaping her life, but I wish she would have just asked my mom for help. To be honest there have been times in my life when I have thought about it, but then I think how hurt my family was when my aunt did it, and I think how I could never put those feelings on anyone. I love the air I breathe, and I'm happy I was blessed with the chance to breathe it. (Craig)

Of course everyone wants to be successful and have all the luxurious things in life, but if I am not going to be happy then what is the use. I could have all the best things in the world but when would I use them if I had to work 80 hour weeks. . . . I want to be successful and still live comfortable and I think that it is very possible to do these things and still work with the things that you love most in life. In my childhood years there is nothing that I have really enjoyed more than waking up to a beautiful day, whatever season it may be. (Gene)

While I am working . . . , I would make sure that I won't be too busy to stay involved with helping the environment and some type of after school program for inner city children. I also hope that I don't become a person that is too concerned with work and doesn't have any time for fun and relaxation. In addition, I plan to make a lot of money, enough to have a nice house and at least two fairly nice cars. The money will be used to open my own landscaping company. . . . Buying a restaurant is another thing I would like to do with my money. . . . Whichever I may choose, my money will definitely be invested into some business of my own. But I also want the money to donate it to foundations that may need it. By this time, I plan to be financially stable and ready to settle with my wife and have two kids. Around this time I should be in my thirties or older because I don't want to get married early. I hope to be set with everything I want and done with going out with the guys on the weekend. (Frank)

In the perfect future I will have a wife who I will have an equal relationship where it will be her choice whether she will work or not. I will have a ten to four job that will support our family sufficiently enough to live comfortably without debt. Children would be nice, three or four, three boys and a girl. [Nick is an only child.] My house would be large but not overdone at all. I would hope that my wife would want to stay home to watch the kids but it would be her decision. I would work in the yard in my free time and play with the kids. My life would be peaceful and stress free. I would sit back and watch my kids grow up and give them a good base for their future and their introduction into the real world. . . . My main goal for the future is to make an impact, not on the world but on my family and friends so that they remember me as a hard working man who loved his family very much. (Nick)

I look into what I hope my future is and I don't see really what I am doing but more what I hope I am not doing. I have many things that I hope I stay away from. My number one rule is to make sure I succeed and avoid failure at all costs. Success has nothing to do with how much money I make or what kind of car I drive. It has nothing to do with my office size, if I even have one, or how many people are below me in a working sense. Success for me means happiness. My fantasy is to raise my kids right. I hope I do all for my wife that I can possibly do. I want to take care of my parents and support them the same way they have supported me. My father told me two things that I will never forget. One is that if you are going to do a job then do it right otherwise don't do it at all. The second thing is that it does not matter what you do in life as long as you do it right. He once told me that he would not care if I was a garbage man just as long as I worked hard at it. I hope to be rich and prosperous, but they are not that important. I would rather be broke and happy than rich and depressed. I will follow my dreams and aspirations. They will lead me to happiness and happiness in life is all any person will ever need. (Harry)

I have given this a lot of thought over the past couple of years, and there are a few things I realize about the career I choose. The most important thing is that I want a career that will give me a good deal of personal satisfaction, a job where I will be working with people directly, making the lives of other people better or more meaningful or happier. I might want to teach, and right now that seems to be the top choice for me. . . . I want to teach because I want to inspire people, young people who are beginning to grab a firm hold of their lives. I not only want to teach the words of literature, but of growing up, and the values by which I stand. To be a friend to them, to help them when they are down and to push them when they are strong. To make others succeed is what makes me feel like I've succeeded, and it is what will give my life much meaning, and an incredible amount of satisfaction. (Brendan)

While I do believe that money does not buy happiness, I sure do think it helps. I would be much happier living in a large house and driving a BMW than I would be if I lived in an apartment and drove a beat-up Chevrolet. No matter how much money I make, I do know that I will be happy because I am going to do something I love to do and believe in. I am not going to do a job that makes me unhappy even if I make a million dollars a year. Well, maybe for a million, but you get the picture. I will always hold a high standard for women. That is not because I am arrogant, but more because I don't want to get involved with someone that does not make me happy. While most guys want a girlfriend because of the idea of having a girlfriend, I want a girlfriend because I truly like the girl, for her beauty, for her personality, for the way she treats me. (Jack)

I have known for a long time that I have an unusual mix of desire and aptitude for fussing with the latest computer technology and a desire to live without computers and in the pure natural world. At some later point in life I have sometimes thought it would be nice to move to some secluded natural area in Maine or Washington state or northern Michigan and live there without day to day dependence on the rest of society. As mentioned in the progression stage this separation of ideals may require a merger where innovation and technology become at one with the natural world. (Shawn)

I imagined a day in my life fifty years from now. I woke up in the morning with my wife (my current girlfriend) at my side. I could hear the birds chirping outside and the sun was shining in through the skylights in the bedroom. "Another beautiful day in paradise, huh, honey," I said to my wife. . . . My wife and I have now been happily married for forty-one years and we have a beautiful and healthy family of five—three boys. We are probably the most happily married couple on the planet. After she and I got up the kids were already downstairs playing the video game system (whichever one is out in 2050) and it was time for breakfast. We only had to program into the computer what we wanted for breakfast and the computer would make the necessary adjustments. . . . Got the boys ready for school, I got dressed and left for work, and my wife went to the plastic surgery office, where she is the Surgeon General. (Byron)

I want to have children one day if the earth is not too overcrowded that laws are passed in the U. S. banning expansion of the population. Assuming I will have the right to produce all the offspring my wife and I want, I want my children to grow up in a place that is not only safe in the way of crime and other ills of human nature, but safe from a tainted environment. I want to raise them in a place similar to [our city] minus the spoiled, drugged out, wannabe nature lovers. Why shouldn't my children grow up in a place that offers the best of human life and nature. I hope that there will be places my wife and I will be able to offer my children that are not tainted with acid rain, pollution, pesticides, lack of forestry, and no safe body of water to be near. Maybe my hopes are idealistic but if others of our generation want such a thing for their children, they need to begin to think about what will make their earth a better place to live. Such a place I would like to go could be near a beach, perhaps one without syringes washing up to the shore. (Clark)

I hope one day to have everything in my life in some sort or another revolve around nature. Whether it is hiking around for exercise or gardening for food, I hope that I am always submersed in Nature. If I have to get a job in the future I would become a guide somewhere. If it is ever time for me to get a career I would go back to school and get a masters and Ph.D. so that I can teach about the thing that has deeply impacted my life. . . . I would like to do some sort of research so that other people can see what living responsibly is like and how easy it really is. (David)

I would like to point out that in saying that I wanted to spoil myself when I got older, I was making a connection with my father. My father honestly has to be one of my biggest role models. He has worked so hard for everything that he has today and still seeks for more. This whole concept fuels me. Coming from Italy at the age of 18, my father hadn't had very much. He came from a farm and put reason in the mentioning of the phrase, "from rags to riches." He has worked so hard to be where he is in life right now, and quite frankly I don't even think he's satisfied yet. I make the connection of spoiling myself with my father because honestly that is how he probably feels in relation to what he had in Italy, which wasn't very much. My father is a hard worker and a perfect example for me to follow for the development of a foundation in life. He's been there, done that. (Greg)

I see myself owning a ranch in central Wyoming. I am a regular cowboy in my dirty brown leather duds. I own a mustang ranch, making my living just enough to support my wife Rachel, and my children Lilly, Chase and Audry [sic]. We each have our own horses, and we ride through the backcountry, going on extended camping trips through the Gros Ventre and Beartooth wilderness. My children and I go hunting occasionally for venison, while my wife, a vegetarian, would eat from the garden we'd grow. Rachel is a teacher at the elementary school in Jackson Hole, an hour away by car. My children would be home taught, learning how to survive both intellectually and physically in the tough surroundings, with help from both Rachel and I. We wouldn't be totally cut off from the rest of the world. We'd have electricity and a TV and other amenities. We'd have a car and a road and work hands to help out on the ranch. We'd take trips all across the country, my family and me, seeing everything worth seeing from the Rio Grande to the rainforest of the Olympic peninsula. We'd see everything, all the while recognizing the rules of environmental impact, i.e. leave only footprints, take only memories, and when you can, don't leave the footprints. It'd be wonderful, we'd ride our horses, and camp and hunt and play with our kids and be totally immersed in the wilderness, all the while making a living. And my standard of success wouldn't be that of most [of this school's] graduates. I wouldn't be making big bucks, but I'd be supporting my family and doing what I loved and being where I loved with the people I love. (Jared)

What this project taught me, ultimately, was that a link with nature, and with life, and with the world, should be in balance and equilibrium with a link with humanity—I believe that if these links remain strong, then I will have the capability of living a "good" life. (Garrison)

Psychiatrist Victor Frankl has written that "man's search for meaning" (the title of one of his books) is the primary motivation in his life (1946). This meaning is idiosyncratic and specific in that it can be fulfilled by him alone (121). Frankl devised a type of psychoanalysis he termed logotherapy to help his patients explore their will to meaning. In explaining logotherapy, Frankl cites a survey taken of almost eight thousand college students run by the social science department at Johns Hopkins University and funded by the National Institute of Mental Health. When the students were asked what they considered "very important" to them, 16 percent checked "making a lot of money;" 78 percent checked "finding a purpose and meaning to my life" (122). Nietzsche says, "He who has a *why* to live for can bear almost any *how*" [emphasis in original] (quoted in Frankl, 126). In the EA, the students, in a small way, were able to explore their will to meaning and begin that search for meaning that would define their lives. As the above student quotes testify, some of them are focused on consumerism, but the vast majority view success as something outside of themselves, where their focus is on family, loving relationships, and a giving back to their community or ecosystem or both.

Schools need to rethink the notion of success. What does it mean to say we graduate successful students? David Orr says:

The plain fact is that the planet does not need more successful people. But it does desperately need more peacemakers, healers, restorers, storytellers, and lovers of every kind. It needs people who live well in their places. It needs people of moral courage willing to join the fight to make the world habitable and humane. And these qualities have little to do with success as our culture has defined it (Orr 1994, 12).

Chapter XIV
Conclusion

> I allow myself eddies of meaning:
> yield to a direction of significance
> running
> like a stream through the geography of my work:
> you can find
> in my sayings
> swerves of action
> like the inlet's cutting edge:
> there are dunes of motion,
> organizations of grass, white sandy paths of remembrance
> in the overall wandering of mirroring mind:
> but the Overall is beyond me: is the sum of these events
> I cannot draw, the ledger I cannot keep, the accounting
> beyond the accounts.
>
> In nature there are few sharp lines: there are areas of
> primrose
> more or less dispersed;
> disorderly orders of bayberry; between the rows
> of dunes
> irregular swamps of reeds,
> though not reeds alone, but grass, bayberry, yarrow, all . . .
> predominantly reeds.
>
> I have reached no conclusions, have created no boundaries,
> shutting out and shutting in separating inside
> from outside: I have
> drawn no line
> as
>
> manifold events of sand
> change the dune's shape that will not be the same
> shape
> tomorrow,
>
> so I am willing to go along, to accept
> the becoming
> thought, to stake off no beginnings or ends,
> establish
> no walls: . . .
>
> —from A. R. Ammon's "Corson's Inlet"

What is *my* definition of success? Has this project been successful? Will I continue in future years in the same vein, without making any major changes to the curriculum?

Conclusion

The first two questions are difficult; I will answer the third in the next paragraph. All I have to go on presently to determine the success of this course is what I see and hear at the very end of these students' high school careers and, from my past years' classes, what responses they may tell me after their first year of college and beyond. The best proof I would have, I feel, is if I saw some activism on the part of my students after they left the school. It need not be a social activism where they get involved in a particular cause, but may instead be a private activism where they themselves make choices that have an ecological component. Even there I cannot be sure it was my course and/or the EA that propelled them in that direction. How is any teacher ever convinced a course was successful? I think a teacher senses whether students are involved and enthusiastic and whether anything the educator has done during a school year has made a significant impression on the student. This is a powerful argument for using naturalistic inquiry for this research (Patton 1990, 13–23). This "sensing" is not quantifiable. No test I give will ever convince me that anything I have done is successful—it is the demeanor and energy of my students that, in the long run, convince me that something has "worked." I have sensed that every year. I used to worry the first year was a fluke, but it was not.

As for future years with this course, I am considering adding another written project to the second semester curriculum. Since my students have had the first semester experience in deconstructing their autobiographies, it seems a shame not to follow up that initial learning with another attempt at deconstruction. We have been deconstructing others' texts, but what about deconstructing another writing of one's own? In reading Noel Gough's essay in Pinar's book *Curriculum: Toward New Identities* (1998), I was attracted to an idea of his. He writes that he has been experimenting with a postmodernist *currere* by using postmodern texts to "diffract the storylines produced by autobiographical writing and personal narrative" (112). He encourages the use of postmodern forms of storytelling—for example, hypertext, metafiction, graphic novels, and cyberpunk science fiction. Questions to be addressed in this ontologically oriented fiction, using these forms, would be: What is a world? How is a world constituted? Are there alternative worlds, and if so, how are they constituted? How do different worlds and different kinds of worlds differ, and what happens when one passes from one world to another? (112). I would also add: How would the world look to a deep ecologist? How would an ecofeminist construct a world? Using ideas of progression, the student would construct a world, and in the construction of that world, he could deconstruct why he was led to that particular scenario.

Could I use this method with my other science classes—chemistry, astronomy, and human anatomy and physiology? Could I see this method being used with other science classes such as physics and biology? Is *currere* a method that has universal application? Does it have application no matter what the age, no matter what the subject? With my limited experience with *currere,* I can only hazard a guess at answers to the above questions.

First of all, I feel *currere* is special. It can't be used for everything; it would lose its luster in commonality. It needs to be carefully orchestrated. It also can't be used for an individual student year after year. There needs to be sizeable space between its applications. For example, I would love to see my students augment their EAs in their final year of college. They would have so many more experiences to contribute, and certainly their perceptions of environment and their place in it would have changed substantially. I foresee an ideal situation in which students would begin their EAs in seventh or eighth grade, add to them in twelfth, and revisit them at the end of undergraduate schooling. This, of course, will never happen; *currere* would need to become a part of universal curriculum, where all, or most, schools would follow this outline, and a student's EA would travel with him from school to school. If *currere* ever became a part of universal curriculum, it would be saying a great deal about the mission of schools—that knowledge of self and knowledge of environment are crucial to a good education.

The widespread use of *currere* would also point up the ecological nature of our own lives. As we examine our pasts and futures and attempt to analyze our reactions to them within the context of both the past and the present, we see the complex interaction of ourselves as biological organisms and the space in which we live. As we, over time, analyze our responses to experiences, we develop new meanings for our lives, which in turn produce new problems, new powers, new demands. There is never a finale; the work is constantly in process.

Secondly, can *currere* be effective with other science courses? Now that I have five years' experience with *currere* in my ecology classes, I am seriously considering trying it with my anatomy classes. Those, too, consist mainly of eighteen-year-olds, and each year I have been teaching that subject, I have been introducing more critical literature into the course. It would be interesting to use the idea of *currere* from a physiological development point. At the regression stage a student would look back at his biological development and see what he can remember about his body's abilities and nuances. In the progression stage, he would look at his body in the future. bell hooks writes often about the inseparability of body and mind, and it would be interesting to explore that in a progression. Analysis and synthesis would focus on how the student perceives himself today physically and mentally, what he has consciously or subconsciously done to get that way, what subtle influences have hegemonically fashioned him, and what conscious decisions he could make to change and/or construct his future.

Conclusion 185

I have tried with very limited success to get my chemistry students (mostly tenth graders) to respond critically to texts they have read. I would have to first explore this area in greater detail before I would think about using *currere* with them, and I would have to think of some chemistry component to carry through the *currere* as a recurrent theme.

I do think *currere* has many applications outside of ecology in high school. It would have to be carefully planned, though, so that one student would not be using the method several times during his high school tenure. But giving the other three hundred or so students in my school who will not be taking ecology during their time at the school the chance to live through the *currere* experience would be so wonderful. The experience is life-altering.

I have taught math and science for many years—biology, chemistry, physics, all manner of math—and have been in a constant state of experiment on the best way to teach these subjects. I believe strongly in an integrated approach to teaching, and I am very much opposed to any teaching where students are merely lectured at and asked to memorize large bodies of vocabulary and equations without much, if any, emphasis on understanding—the banking approach as Freire describes it (1970, 52–67), where students are the depositories and the teachers are the depositors. Many others besides me have been concerned about the lack of scientific and mathematical conceptual understanding among students, especially at the elementary and high school level, and over the years countless "recipes" have been proposed to deal with these issues. No matter how many conferences I attend, how many classes I take, how many educators I talk with, I continue to see this frustration in the teaching of science. The more I read of Pinar, the more I feel we may be in need of a "reconceptualization" of the mainstream teaching of science, a Kuhnian idea of revolutionary science (Kuhn 1962). Most students who take science courses in high school will not go on to be scientists, just as most algebra students will not go on to be mathematicians. How do we get the majority of science-course-taking students invested in science? How do we move away from this elitist posturing so apparent in traditional methodology of science teaching where science must be "hard" and difficult, a place where a weeding-out occurs as well as a "turning off"? I don't want students turned off to science. Science is much too interesting, and the issues studied, whether in ecology or chemistry, directly affect people's lives. *Currere*, used in the way addressed in this book, may prove to be a new paradigm for getting students invested in science.

There is power in *currere*, a power that is social as one goes about the four stages of the process. In *currere*'s adaptation in the EA, that power goes one step further—it is power that goes toward creating a better world. As the student becomes aware of how important environment is in his own life, he becomes connected to the space in which he has lived and in which he hopes to live, and the seeds of stewardship are sown.

In May 2000, at the end of the second year of this experimental course, I run into a student from the first year's class. He has had a great first year at college and is looking forward to being home for the summer and making some money. He tells me he has managed to check off two things on his Life List. He may be "jiving" me, as students are wont to do, but as he is leaving he says, "You know, I want to use some of this downtime this summer to add to my EA; I have so many things I want to write about my college experiences and so much I want to record about the environment on my campus."

To paraphrase Pinar: Mind in its place, I conceptualize that present situation. I choke out a "so long" to him and wish him a good summer.

I am placed together.

Synthesis.

Appendix
How I Did My Research

When I left Penn State after my year of residency, my coursework behind me, I returned to my full-time teaching job. I had been given only one year's leave from my school and also was in no position to live another year without any income. I left Penn State, though, excited about beginning my dissertation and also excited about getting back to working with students—the one thing I had really missed in my otherwise amazing experience at Penn State—and employing some of the new material I had gleaned from my doctoral studies. I had a topic for my dissertation and had already put together two different proposals for that topic, unsure as to how best to approach that idea. I made a schedule for myself as to when I would work on my schoolwork and when I would work on my dissertation, and I strongly felt that in two years' time, I would have the dissertation finished.

By the end of September, I was awash in work for my classes back in Cleveland and had done nothing on my dissertation. Part of the problem stemmed from the fact that I had three preparations every day, one of those in a subject that had never been taught at my school, and a second in a subject that had not been taught for over seven years. Not only did I have to write curriculum, but I also had to create, prepare, and organize labs. I felt inundated with work. By the end of October, I was depressed; my workload was continuing to mount and I had still done almost nothing on my dissertation. The one bright spot in my day, every day, was my ecology classes. I couldn't get over how well they were going. By the middle of November, I had decided to use this huge amount of schoolwork to my advantage. I redirected my ideas for the dissertation (swearing to myself that after this project I would get back to my original idea, a project I still very much want to pursue). Once I began viewing my exhaustion at the end of the day as a positive loss of energy, I began to see the impact of this new dissertation idea.

I had been religiously taking notes every week on all my classes from the first day in late August. Each Sunday I would sit at my computer and type notes on how the three classes had gone the previous week (I was then teaching ecology, human anatomy and physiology, and chemistry). I wanted to do this so I would have good notes for subsequent years, as I was literally building up new files on two of these three subjects. I realized that in doing this I had already started taking field notes for my dissertation. Once I made the change to this new topic, I began to write these field notes with even more detail, following Patton's algorithm for effective field notes (Patton

1990, 239–44).[1] I also began photocopying students' work, not only of the beginning stages of the EA but also any essays or observations they would make that I thought particularly pertinent to my new dissertation topic.

In December I met with the then head of my committee, Dr. Jeanne Brady, at Penn State to get her thoughts on this new idea and to ask her to help me organize my approach. Together we decided that there needed to be a critical pedagogy piece, a feminist piece, and a historical piece on the teaching of ecology. We discussed Sandra Harding and Donna Haraway, two women writing about new approaches to the teaching of science, and how their work could help frame my argument. I left Penn State that Saturday afternoon reinvigorated and with a workable time frame in which to continue working at my job and still get a dissertation researched and written. Thank you, Jeannie, for your encouragement and confidence in my ability to see this project to a conclusion. At that point in the project, I was not at all sure I could pull it off.[2]

Come January and the end of the first semester and the EA projects were finished. I photocopied each student's EA and at that point made copious notes to myself about where I felt that student had gone with the EA. Throughout the second semester I maintained my field notes, made further photocopies of pertinent student work, and continued reading in the fields of critical pedagogy, feminist epistemology, and the teaching of science.

I was also busy preparing to take my comprehensives in June. With the help of Dr. Brady and the rest of my committee, we came up with topics for essays I would have to write as part of the comprehensives that had to do with my dissertation study. I worked on these most every weekend from January to June. The essays helped me to constantly refocus on the bigger picture of my ecology classes. During the school week I was bogged down in the daily routines of teaching, but on the weekend, the work on the comprehensives forced me to step back and look at the big picture of where the class was going. Did something that happened that week come about because of the EA project? Were students growing in the class? What about the synergy

---

[1] Probably the most important skill a researcher can have is observation. Virginia Woolf, not a researcher, realized the importance of observation for her writing: "Odd how the creative power at once brings the whole universe to order. . . . I mark Henry James' sentence: observe perpetually. Observe the oncome of age. Observe greed. Observe my own despondency. By that means it becomes serviceable."

[2] Dr. Brady has since gone on to teach at St. Joseph's outside of Philadelphia. She has continued to advise me throughout this process, but was unable to stay on as head of my committee when she left Penn State in June 1999. I will always be grateful to her for many things, but I think particularly for her course in feminist history and epistemology, in which I literally reformulated for myself where I want to go as a feminist. She also introduced me to a wonderful group of fellow students in that class who continued to be some of my strongest allies and supporters throughout this project.

and symbiosis that appeared to be developing within the two sections? This change in outlook allowed me to see much more than I would normally have seen. For example, I learned in January that I was to have a new student in one of my ecology sections. Normally, I would have shrugged my shoulders, said okay, and not thought much more about it. But this time, I saw a new student (in fact, one I already knew) entering the class as an interesting experiment. He would not have written an EA; he would be coming in as an outsider into this section that had definitely bonded and grown together, and he had had no ecology class prior to this change. He would become a control, someone I could observe along with my yearlong students to see if the EA did indeed make a difference in the conduct of the class.

Once the comprehensives were over, I used all of July and the first two weeks of August to concentrate on my first-year EAs. I reread them all and started looking at different ways I could organize their data. Did I want to group together sections that had particular themes, a cross-case analysis? Or did I want to take each EA idiosyncratically and view that against a background of the student? What were students saying in their choice of language, and what were they saying by their absence of certain words and ideas in their data? I was attempting a hermeneutical analysis that I also tried to historicize. What was going on in the student's life ten, eighteen years earlier that could have added to his perception of what he wrote? What hegemonic tendencies were at work in the student's life that were evident in his writing?

The third week of August I had to get back to my normal schoolwork and prepare for my second round of ecology classes. I continued to keep field notes and make photocopies of pertinent student work. I was careful to stay as close as possible to the techniques I had used the previous year in presenting the material, especially the EA, as I didn't want to influence the outcome of the project in any other way than I had the year before. I did make some changes in the syllabus for the course—a pilot mapmaking project, a unit on the politics of billboards, a better-organized approach to the EE, a more frequent use of the critical literacy criteria for the students' essay writings, an introduction to "green buildings" and the idea of sustainable architecture—but I kept the introduction of new sections of the EA and all the follow-up and rough draft sessions exactly as I had done them previously.

During the early fall, I was reorganizing my dissertation committee. I had to find a new head to replace Dr. Brady. I eventually managed to convince one of my original committee members, Dr. Dan Marshall, to take on the role of head, and found a new committee member from the Department of Curriculum and Instruction, Dr. Jim Nolan. In November, I met with my new committee to go over my proposal and look at my first several chapters. We had a session in which we worked on a revised outline for the paper and discussed in detail what I wanted to have at the center of my study.

During the fall months I continued to read in the field and reevaluate the previous week on weekends, as I had done the first year. These evaluations didn't take as long as the first year, so I had more time to work on the actual dissertation writing. Dr. Marshall had also made it possible for me to contact William Pinar via cyberspace. I sent him the early chapters I had brought with me to the November committee meeting, and Dr. Pinar wrote me back copious notes on my writings. Dr. Pinar also agreed to be a special signatory on my committee. He sent me a copy of *Toward a Poor Curriculum,* his 1976 book in which he first outlined his thoughts for *currere*. I had been unable to find this book and had asked him if he had any ideas where I might go to locate one. A week later, I found one in my mailbox.

In January 2000, I received a new collection of data: thirty EAs that I needed to read and reread in order to add to my first year's data.

The value of doing a classroom research is not to be underestimated. When writing the proposal(s) for my first dissertation idea, I had to think about how I would get my interviewees to open up to me, to trust me, how to best question them so that I was getting the data I was looking for. I had to think about a setting for this questioning. Should it be in a classroom, a neutral place, a place outside their normal workplace? How should I configure the space? I did not have to deal with any of these issues in a classroom research. I was the teacher, I controlled the space from the first day of classes. My data sources were comfortable with the space, used to it, able to make small changes to the space to suit them. I also knew my data sources. I was teaching some of them for a second and third time. I did not have to worry about making my time with them productive for the dissertation because I had many hours with them, not just a few interviewing sessions. Seeing them daily for nine months also allowed for periods of time when something unexpected would happen. Because of the small size of our school, I also knew many of these students' parents, which gave me further information I could use when analyzing my students' data. The students were used to me, and there was nothing contrived in their reactions (except, of course, they were perfectly aware that I was grading them). I not only worked with the students on the EA in class but also conducted one-on-one sessions at different stages of the project, where I could concentrate on one student's writings and help stretch him to move more deeply into the experience. I was not the disembodied Other, directing and documenting the research; I was a collaborative participant, working with and encouraging my students.

Qualitative researchers often write about the need for group processing—finding someone to discuss research with and get feedback on analysis. I have to say I was pretty much unable to do this. No faculty member in my science department was even remotely interested in what I was doing. Because I was employing something other than scientific method—although I was using scientific method for other aspects of the class—I was considered to be on the fringe. I did contact Dr. David Orr, head of the environmental sciences department at Oberlin College, and he was supportive, but other than that, I was alone. I would not advise anyone to work for two or more years by her/himself without the type of support I am referring to here. I received wonderful support from my committee members, especially the head of my committee, Dr. Dan Marshall, and from Dr. Pinar, but I often found myself sorely missing an ecologist or other scientist with whom to discuss my work. One member of my committee, Dr. Tom Martin, is a professor in the Department of Wildlife and Fisheries at Penn State. He always seemed very open to what I was doing and curious about how the project was going, but I never felt he would try something like this himself or recommend the method to someone else. He is highly gifted in statistics and I know felt it his duty to make sure his students, especially his doctoral students, were able to publish their researches using clear, accurate statistical methods. He also was and is extremely busy and really had little time in his life for a qualitative research. Dr. Jim Nolan, the newest member of my committee, a professor in the Department of Curriculum and Instruction, was very receptive to the techniques I was trying, and it was he who in a full committee meeting in December 1999 said something that I have not been able to keep out of my mind: "Don't let the voice of the students get lost." At the time I was working with an outline similar to Paul Willis's organization in *Learning to Labor,* where he divides the book into two parts: the first part being his collection of data, the second part his analysis of that data. If I continued with that organization, Dr. Nolan was right—the voices of the students would get lost. From that meeting, I was determined to find an organization for the dissertation that would focus on the voices of the students.

When I first decided to go ahead with this idea of a classroom research, I struggled with calling it a case study approach or a phenomenological study. I ended up making an amalgam of both approaches. A classroom research in which the researcher has provided the curriculum being analyzed becomes an intense, personal experience. The culture of the school in which I teach, the culture of science teaching in general, and the culture of teenagers are all cultures in which I am deeply involved. Because of this, I see the theoretical framework guiding my research to be heuristic phenomenology as seen through case study (Moustakas 1990). It is heuristic because of my personal experience and because I have an intense interest in how this curriculum is perceived.[3] It is phenomenological because there is a structure and essence

---

[3] Merriam describes the characteristics of a heuristic case study: It can explain why an innovation worked or failed; discuss alternatives not chosen; and evaluate, summarize, and conclude, increasing its potential applicability (Merriam 1998, 31). This certainly was true in my study.

not only to the culture of a high school class, but also to the unique aspect of a shared experience of *currere* (Patton 1990, 69–73). The project is a case study because the experience involves one teacher teaching at one school using one curriculum (Merriam 1998, 29–33).

Heuristic inquiry can be divided into five steps: immersion, incubation, illumination, explication, and creative synthesis (Patton 1990, 73). I, the researcher, am the primary instrument in these steps as I immerse myself in observation and hermeneutics in which, through periods of incubation, I can use my own personal experience to elucidate the experiences of others to come to some disclosure and explication of what was happening in the classroom through this process of *currere*. Heuristic inquiry, Patton declares, "challenges in the extreme tradition scientific concerns about researcher objectivity and detachment" (73).

Peter Reason uses the term "critical subjectivity" to refer to a quality of awareness in which we do not suppress our primary experience, nor do we allow ourselves to be swept away and overwhelmed by it; rather we raise it to consciousness and use it as part of the inquiry process (Reason 1988, 12). This subjectivity, I feel, is important to my research. There are some things that I know from my experience in this culture will give me solid grounding for this work. What I need to do is critically look at these "things I know" and analyze the meanings I find for myself in these "knowings."

Before I began my doctoral work at Penn State, I knew I wanted to do a qualitative research. I had had plenty of opportunities to do quantitative researches in science, and I had developed a bias against them, having seen too many times how statistics could be manipulated. I knew there were inherent problems with qualitative researches also (Patton 1990, 13–14), but I wanted to try one of my own. As I continued to gather data for this particular dissertation, I could see how much rich, thick data I was accumulating. Not only did I have the advantage of participant observation, but I also was acquiring pages and pages of text from each student that I could analyze in many different ways. How the students understood the project, how they saw themselves as part of an ecosystem, an ecosystem for which they have a degree of responsibility, was contextual; the reality of their experiences was constructed by these participants in their interaction with me, with their peers, with their families and friends, and with their knowledge of the world in general. I could not possibly quantify that into some statistical program.

My sampling was nonrandom and purposeful (Bogdan and Biklen 1998, 65). I was using my own students, a group not necessarily selected by me ahead of time, but some of the students signed on to the course because I was

---

Something had happened to my ecology classes that first year because of *currere*. I wanted, needed to understand what that was.

teaching it. I was using the data for a distinct purpose, not simply reading the data to see where it would lead me, but analyzing it to see if I could understand (1) how the process of *currere* so engaged the students and (2) how successful the course was because of this core project.

As I analyzed the Eas, I was asking myself these questions: (1) How has the writing of the EA affected the student? (2) Does the student see himself differently at the end of the project? (3) Has the student related any particularly telling event that has helped form who he is today? (4) Has the sharing of the EA work in the classroom with his peers helped or hindered his own process? (5) Do I, as classroom teacher, find any direct link between this student's EA and the rest of his work in the course?

I also made a conscious decision not to let my students know their work was going to be part of a research project. I didn't want anything to interfere with the writing process. By the time we were ready to begin the EA, the students were used to me as their teacher, and there was nothing contrived in their reactions to the EA or to the one-on-one sessions we would have at the different stages of the EA. The first year's students were aware that I was engaged in research in general (I was also at that time working on a book on SAT testing that I had to get finished by March, and they sometimes heard me talking about it). It was only after we finished the EA project that I asked them how they would feel about my using their work as part of my research. I really didn't even have to broach the subject because on two different prior occasions students had suggested that I publish their writings, saying it would be in the readings of those writings that people might finally understand what it is like to be a teenager. When I specifically asked the class about using their data, not only was there total approbation for the idea, but many students wanted me to use their entire EA in whatever I was going to publish. Their excitement at this point further convinced me I was right in not telling them ahead of time about their being involved in a research project.

For the analysis, I decided on a modified analytical induction approach (Merriam 1998, 160) where, after early data collection, I formed a hypothesis that the EA was making my students more enthusiastic about the field of ecology and more invested in the science. Then with subsequent data, I looked to negate my working hypothesis (Popper would approve [Popper 1934]!) Purposeful sampling is a necessary component of analytic induction, because in the purposeful sampling the researcher chooses subjects that are believed to facilitate the expansion of the developing theory (Bogdan and Biklen 1998, 65). In this modified version of analytical induction, the researcher considers a rough explanation of the observed phenomenon early on in the process, then holds up new data as it is collected to that rough explanation. If, at this point, the explanation needs to be modified to fit the new data, that is done, but the researcher is always on the lookout for data that does not match the existing explanation. This process continues throughout the data collection (Bogdan and Biklen 1998, 65). I never once found a case, in the more than 135 cases I studied, that did not meet my original working hypothesis: the EA *was* making my students more enthusiastic about ecology and more invested in the science.

Since I am the one who had constructed this curriculum, I had to be on guard for the potential of bias in my analysis. I had to constantly remind myself to be undetermined and nonleading during the data collection and interpretation stages (Bogdan and Biklen 1998, 33–34). Merriam talks about the idea of *epoche*, the process whereby the researcher engages to remove (or at least become aware of) prejudices, viewpoints, or assumptions regarding the phenomenon under investigation (Merriam 1998, 158). It is a suspension of judgment and an awareness of judgment. The researcher sets aside personal viewpoint in order to see the experience for itself. It is also important to see the experience from several different perspectives, what Moustakas calls "imaginative variation" (Moustakas 1990). Using imagination, the researcher varies frames of reference in order to arrive at contextual descriptions of experience. Since I am building my conceptual framework on heuristic phenomenology, I also use my own reflective thoughts as part of my data. Moustakas writes:

> The task of imaginative variation is to seek possible meanings through the utilization of imagination, varying the frames of reference, employing polarities and reversals, and approaching the phenomenon from divergent perspectives, different positions, roles, or functions. The aim is to arrive at structural descriptions of an experience, the underlying and precipitating factors that account for what is being experienced. How did the experience of the phenomenon come to be what it is? (97–98)

I have kept an audit trail to answer questions about consistency of results and verifiability (Erlandson et al. 1993, 34–35). In doing naturalistic inquiry, the researcher knows that objectivity is an illusion and that no methodology can be divorced from the researcher who has selected and used it. I cannot assure that my observations are free from contamination of my own prejudices, and so I have literally thousands of pages of data in my audit trail to help confirm my conclusions.

The fact that I have five school years of data goes a long way, I think, in helping me and my readers decide how successful I think the course is and, in particular, how integral the *currere* process is to that success or lack of success. Without those five years of data, I could be looking at a fluke, a blip on the radar screen. I presented data from many of my students so that I could not be accused of choosing only data that reinforced my hypothesis. I also chose to present that data in a random fashion, deliberately not choosing to present the "best" data at the end. In fact, I made no attempt to rank the data in a cross-case analysis, cross-class analysis, or cross-year analysis. I did choose to present the five years' data in two separate chapters, in order to focus on my concerns as the practitioner of finding that same sense of energy and engagement that I experienced in the first year. Rosanna Hertz, in her introduction to *Reflexivity and Voice* (1997), talks about issues of voice—how authors express themselves in their work:

> The respondents' voice is almost always filtered through the author's account. Authors decide whose stories (and quotes) to display and whose to ignore. The decision to privilege some accounts over others is made while developing theories out of the data collected. As they shift between data and theory, scholars make decisions about the voices and placement of respondents within the text. (xii)

By choosing to represent many of my students with their own words, although contextualized by me, I hoped to limit my voice, and instead do as Dr. Nolan suggested: concentrate on the voice of my students.

How generalizable are the results of this work? "Generalizability" is not a word that sits well in the heart of the naturalistic researcher (Kincheloe 1991, 127–42; Merriam 1998, 207–13). Context shifts over time; all observations are defined by the specific context in which they take place (Erlandson et al. 1993, 32). Since I am using a modified analytical induction approach, where after early data collection I formed a hypothesis that the EA was making my students more enthusiastic about the field of ecology and more invested in science, I was always looking to negate my working hypothesis. As Patton suggests, I was, and am, looking to provide perspective, not truth, an "empirical assessment of local decision makers' theories of action rather than generation and verification of universal theories, and context-bound extrapolations rather than generalizations" (Patton 1990, 489). Guba and Lincoln agree there is a fundamental distinction concerning generalizability in traditional versus naturalistic researches. In a traditional study it is the obligation of the researcher to ensure that findings can be generalized to the population; in a naturalistic inquiry it is the obligation of the readers, those who would apply the study to the receiving context, to show transferability (Guba and Lincoln 1985, 241). In providing rich, thick description of the process involved and the data collected, I hope to allow the reader to decide whether her/his context fits with mine.

BIBLIOGRAPHY

American Association for the Advancement of Science. (1993). *Benchmarks for science literacy: Project 2061*. New York: Oxford University Press.

American Chemical Society. (1996). *ChemCom: Chemistry in the community*. (3rd ed.). Dubuque, IA: Kendall/Hunt Publishing Co.

American Institutes for Research. (1999). *An educator's guide to school–wide reform*. Arlington, VA: Educational Research Service.

Barrett, A. (1998). *The voyage of the* Narwhal. New York: W. W. Norton & Co.

Berger, A. A. (1995). *Cultural criticism: A primer of key concepts*. Vol. 4. Thousand Oaks, CA: Sage Publications.

Bogdan, R. C., & Biklen, S. K. (1998). *Qualitative research in education: An introduction to theory and methods*. (3rd ed.). Boston: Allyn and Bacon.

Burke, M. M. (1984). *Reciprocity of perspectives: An application of the work of James B. McDonald to a personal perspective of special education (curriculum theory)*. Unpublished doctoral dissertation, University of North Carolina, Greensboro.

Campbell, N. A., Reese, J. B., & Mitchell, L. E. (2000). *AP biology*. Menlo Park, CA: Addison-Wesley.

Camus, A. (1948). *The plague*. New York: New York Modern Library, 1989.

Carson, R. (1962). *Silent spring*. Boston: Houghton Mifflin.

Chase, A. (2000). Harvard and the making of the Unabomber. *The Atlantic Monthly* 285, June, 41–65.

Chodorow, N. (1978). *The reproduction of mothering*. Berkeley: University of California Press.

Clark, M. (1999). Wounds, legal battles still fester in Alaska. *The Plain Dealer,* March 21, 5-H.

Cole, N. J. (1984). *A descriptive study of an elementary school.* Unpublished doctoral dissertation, Oklahoma State University.

Connelly, M., & Clandinin, J. (1990). Stories of experience and narrative inquiry. *Educational Researcher* 19(4), 2.14.

Daniel, D. K. (1991). *Teaching as hermeneutics.* Unpublished doctoral dissertation, University of Alberta, Canada.

Davies, P. C. W. (1988). *The cosmic blueprint: New discoveries in nature's creative ability to order the universe.* New York: Simon & Schuster

Didion, J. (1961). *Slouching towards Bethlehem.* New York: Farrar, Straus and Giroux.

Djerassi, C. (1994). *The Bourbaki gambit.* New York: Penguin Books.

Educational Testing Service (1999). *Environmental Science Advanced Placement course description.* Princeton, NJ: College Entrance Examination Board and Educational Testing Service.

Eisenhart, M. (1996). The production of biologists at school and work: Making scientists, conservationists, or flowery bone-heads? In B. Levinson, D. Foley and D. Holland (Eds.), *The cultural production of the educated person: Critical ethnographies of schooling and local practices*, pp. 169-85. Albany, NY: State University of New York Press.

Erlandson,, D. A. et al. (1993). *Doing naturalistic inquiry: A guide to methods.* Thousand Oaks, CA: Sage.

Ertel, E. W., & Scoll, P. D. (1997). *Teacher's guide: AP environmental science.* Princeton, NJ: The College Entrance Examination Board and the Educational Testing Service.

Exxon Corporation. (1992). *Scientists and the Alaska oil spill: The wildlife, the cleanup, the outlook.* [video]. Exxon, USA.

Feinberg, P. R. (1982). *A Buberian critique of four curriculum theorists.* Unpublished doctoral dissertation, Loyola University, Chicago.

Foucault, M. (1980). *Power-Knowledge: Selected interviews and other writings, 1972-1977.* New York: Pantheon.

Fox, W. (1995). The deep ecology–ecofeminism debate and its parallels. In G. Sessions (Ed.), *Deep ecology for the 21st century* (pp. 269–289). Boston: Shambhala Publications.

Frankl, V. E. (1946). *Man's search for meaning.* New York: Washington Square Books, 1984.

Freire, P. (1970). *Pedagogy of the oppressed.* (20th anniversary ed.). New York: Continuum.

Freire, P. (1992). *Pedagogy of hope.* New York: Continuum.

Freire, P. (1996). *Letters to Cristina: Reflections on my life and work.* New York: Routledge.

Gallagher, W. (1993). *The power of place: How our surroundings shape our thoughts, emotions, and actions.* New York: HarperCollins.

Gilligan, C. (1982). *In a different voice: Psychological theory and women's development.* Cambridge: Harvard University Press.

Giroux, H. A. (1981). *Ideology, culture, and the process of schooling.* Philadelphia: Temple University Press.

Giroux, H. A. (1997). *Pedagogy and the politics of hope: Theory, culture, and schooling.* Boulder, CO: Westview Press.

Gleick, J. (1987). *Chaos: Making a new science.* New York: Penguin Books.

Goldberger, N., et al. (1996). *Knowledge, differences, and power: Essays inspired by women's ways of knowing.* New York: Harper Collins.

Gore, A. (1992). *Earth in the balance: Ecology and the human spirit.* New York: Houghton Mifflin.

Gottfried, S. S. (1993). *Biology today.* St. Louis: Mosby Publications.

Gough, A. (1998). Beyond Eurocentrism in science education: Promises and problematics from a feminist poststructuralist perspective. In W. F. Pinar (Ed.), *Curriculum: Toward new identities* (pp. 185–210). New York: Garland Publishing.

Gough, N. (1998). Reflections and diffractions: Functions of fiction in curriculum inquiry. In W. F. Pinar (Ed.), *Curriculum: Toward new identities* (pp. 93–128). New York: Garland Publishing.

Graham, R. J. (1992). Currere and reconceptualism: The progress of the pilgrimage 1975–1990. *Journal of Curriculum Studies* 24(1, January–February), 27–42.

Greene, M. (1984). *Landscapes of learning.* New York: Teachers College Press.

Gress, J. R., & Purpel, D. E. (Eds.). (1988). *Curriculum: An introduction to the field.* Berkeley, CA: McCutchan.

Grumet, M. R. (1975a, April). *Existential and phenomenological foundations of currere: Self-report in curriculum inquiry.* Paper presented at the American Educational Research Association annual meeting, Washington, D.C.

Grumet, M. R. (1975b, April). *Supervision and situation: A methodology of self-report for teacher education.* Paper presented at the American Educational Research Association annual meeting, Toronto, Canada.

Grumet, M. R. (1981). Restitution and reconstruction of educational experience: An autobiographical method for curriculum theory. In M. Lawn & L. Barton (Eds.), *Rethinking curriculum studies: A radical approach* (pp. 115–30). London: Croom Helm.

Grumet, M. R. (1988). *Bitter milk: Women and teaching.* Amherst: University of Massachusetts Press.

Grumet, M. R. (1992). Existential and phenomenological foundations of autobiographical method. In W. F. Pinar & W. M. Reynolds (Eds.), *Understanding curriculum as phenomenological and deconstructed text* (pp. 28–43). New York: Teachers College Press.

Guba, E. G., & Lincoln, Y. S. (1985). *Naturalistic inquiry.* Newbury Park, CA: Sage.

Harding, S. (1991). *Whose science? Whose knowledge?* Ithaca, NY: Cornell University Press.

Hartslee, R. J. (1999). *Just as we are: Education, experience and William Pinar's poor curriculum*. Unpublished doctoral dissertation, University of North Carolina, Greensboro.

Hertz, R. (Ed.). (1997). *Reflexivity and voice*. Thousand Oaks, CA: Sage.

Hill, J. B. (2000). *Legacy of luna: The story of a tree, a woman, and the struggle to save the redwoods*. New York: Harper Collins.

hooks, b. (1994). *Teaching to transgress: Education as the practice of freedom*. New York: Routledge.

Huebner, D. (1974). Toward a remaking of curricular language. In W. F. Pinar (Ed.), *Heightened consciousness, cultural revolution, and curriculum theory* (pp. 36–53). Berkeley, CA: McCutchan.

Huebner, D. (1975). Curriculum as concern for man's temporality. In W. F. Pinar (Ed.), *Curriculum theorizing: The reconceptualists*. Berkeley, CA: McCutchan.

Hyder, T. W. (1998). *A model for remedial writing instruction based on autobiographical theory and informed by critical pedagogy (community colleges, composition, writing instruction)*. Unpublished doctoral dissertation, University of Virginia, Charlotte.

Ingalls, Z. (2000). Green building at Oberlin is a new dream house for environmental studies. *Chronicle of Higher Education*, January 21, B2.

Janovy Jr., J. (1985). *On becoming a biologist*. Lincoln: University of Nebraska Press.

Kincheloe, J. L. (1991). *Teachers as researchers: Qualitative inquiry as a path to empowerment*. Vol. 4. New York: The Falmer Press.

Kincheloe, J. L. (1993). *Toward a critical politics of teacher thinking: Mapping the postmodern*. Westport, CT: Bergin & Garvey.

Kincheloe, J. L. (1998). Pinar's currere and identity in hyperreality: Grounding the post-formal notion of intrapersonal intelligence. In W. F. Pinar (Ed.), *Curriculum: Toward new identities* (pp. 129–42). New York: Garland.

King, Y. (Ed.). (1994). *Dangerous intersections: Feminist perspectives on population, environment, and development*. London: South End Press.

Kliebard, H. M. (1995). *The struggle for the American curriculum 1893–1958*. (2nd ed.). New York: Routledge.

Kuhn, T. S. (1962). *The structure of scientific revolutions*. (3rd ed.). Chicago: University of Chicago Press.

Lather, P. (1991). *Getting smart: Feminist research and pedagogy with/in the postmodern*. New York: Routledge.

Lear, L. (1997). *Rachel Carson, witness for nature*. New York: Henry Holt and Company.

LeCompte, M. D. (1995). Some notes on power, agenda, and voice: A researcher's personal evolution toward critical collaborative research. In P. L. McLaren & J. M. Giarelli (Eds.), *Critical theory and educational research* (pp. 91–112). Albany: State University of New York Press.

Leopold, A. (1949). *A Sand County almanac*. New York: Ballantine, 1966.

Levant, R. F., & Pollack, W. S. (Eds.). (1995). *A new psychology of men*. New York: Basic Books.

Lincoln, Y. S. (1992). Curriculum studies and the traditions of inquiry: The humanistic tradition. In P. W. Jackson (Ed.), *Handbook of research on curriculum* (pp. 79–98). New York: Macmillan.

Lopez, B. (1989). The American geographies. *Orion Nature Quarterly* 8(4), Autumn, 50–62.

Lovelock, J. E. (1979). *Gaia, a new look at life on earth*. Cambridge: Oxford University Press.

Lusted, D. (1986). Why pedagogy? *Screen* 27(5), 2–14.

Lyons, N. (1983). Two perspectives on self, relationships and morality. *Harvard Educational Review* 53, 125-45.

Mader, S. (1998). *Human biology*. Boston: McGraw-Hill.

Marshall, J. D., Sears, J. T., & Schubert, W. H. (2000). *Turning points in curriculum: A contemporary American memoir.* Upper Saddle River, NJ: Prentice-Hall.

Matthiessen, P. (2000). Rachel Carson, Time Magazine, millennium issue, 190.

Maxwell, J. A. (1996). *Qualitative research design: An interactive approach.* (Vol. 41). Thousand Oaks, CA: Sage Publications.

McKibben, B. (1989). *The end of nature.* New York: Doubleday.

McKibben, B. (1998). A special moment in history. *The Atlantic Monthly* 281, May, 55–78.

McLaren, P. (1989). *Life in schools: An introduction to critical pedagogy in the foundations of education.* New York: Longman.

Merleau-Ponty, M. (1962). *Phenomenology of perception.* London: Routledge.

Merriam, S. B. (1998). *Qualitative research and case study applications in education.* San Francisco: Jossey-Bass Publishers.

Merton, T. (1985). *Love and living.* New York: Harcourt Brace Jovanovich.

Miller, G. T. (1991). *Environmental Science* 3rd edition. Belmont, CA: Wadsworth Publishing Co.

Miller, J. L. (1990). *Creating spaces and finding voices: Teachers collaborating for empowerment.* Albany: State University of New York Press.

Miller, J. (1997). *Egotopia: Narcissism and the new American landscape.* Tuscaloosa: University of Alabama Press.

Mohanty, J. N. (1997). *Phenomenology: Between essentialism and transcendental philosophy.* Evanston, IL: Northwestern University Press.

Montagu, A. (1997). Foreword. In J. Miller, *Egotopia: Narcissism and the new American landscape* (pp. ix–xiii). Tuscaloosa: University of Alabama Press.

Montessori, M. (1917). *Spontaneous activity in education: The advanced Montessori method*. New York: Schocken Books (1965).

Montessori, M. (1949). *Education and peace*. Chicago: Henry Regnery Co.

Morgan, W. (1997). *Critical literacy in the classroom: The art of the possible*. New York: Routledge.

Moustakas, C. (1990). *Heuristic research: Design, methodology, and applications*. Thousand Oaks, CA: Sage.

Myers, C. F. (1983). *A personal inquiry, through currere, into the person/earth relationship, using the hermeneutic spiral as model*. Unpublished doctoral dissertation, University of North Carolina, Greensboro.

Mykhalovskiy, E. (1997). Reconsidering "Table talk": Critical thoughts on the relationship between sociology, autobiography, and self-indulgence. In R. Hertz (Ed.), *Reflexivity and voice* (pp. 229–251). Thousand Oaks, CA: Sage Publications.

Noddings, N. (1992). *The challenge to care in schools*. New York: Teachers College Press.

Odum, E. P. (1971). *Fundamentals of ecology*. (3rd ed.). Philadelphia: W. B. Saunders Co.

Orr, D. W. (1992). *Ecological literacy: Education and the transition to a postmodern world*. Albany: State University of New York Press.

Orr, D. W. (1994). *Earth in mind: On education, environment, and the human prospect*. Washington, DC: Island Press.

Orr, D. W. (1998). So that all the other struggles may go on. *The Observer*, January 30, 4.

Patton, M. Q. (1990). *Qualitative evaluation and research methods*. (2nd ed.). Newbury Park, CA: Sage Publications.

Payne, M. (1997). *A dictionary of cultural and critical theory*. Oxford, UK: Blackwell Publishers.

Pinar, W. F. (1974a). Currere: Towards reconceptualization. In J. Jelinek

(Ed.), *Basic problems in modern education* (pp. 147–171). Tempe: Arizona State University, College of Education.

Pinar, W. F. E. (1974b). *Heightened consciousness, cultural revolution, and curriculum theory: The proceedings of the Rochester conference.* Berkeley, CA: McCutchan.

Pinar, W. F. (1975a). Currere: Towards reconceptualization. In W. F. Pinar (Ed.), *Curriculum theorizing: The reconceptualists* (pp. 396–414). Berkeley, CA: McCutchan.

Pinar, W. F. (1975b). *The method of currere.* Paper presented at the American Educational Research Association annual meeting, Washington, D.C.

Pinar, W. F. (Ed.). (1975c). *Curriculum theorizing: The reconceptualists.* Berkeley, CA: McCutchan Publishing.

Pinar, W. F. (1978a). Currere: A case study. In G. Willis (Ed.), *Qualitative evaluation* (pp. 316–42). Berkeley, CA: McCutchan.

Pinar, W. F. (1978b). Life history and curriculum theorizing. *Review Journal of Philosophy and Social Science* 3(1), 92–118.

Pinar, W. F. (1994). *Autobiography, politics, and sexuality: Essays in curriculum theory, 1972–1992.* New York: Peter Lang.

Pinar, W. F., ed. (1998). *Curriculum: Toward new identities.* New York: Garland Publishing.

Pinar, W. F., & Grumet, M. R. (1976). *Toward a poor curriculum.* Dubuque, IA: Kendall/Hunt.

Pinar, W. F., Reynolds, W. M., Slattery, P., & Taubman, P. M. (1995). *Understanding curriculum: An introduction to the study of historical and contemporary curriculum discourse.* New York: Peter Lang.

Popper, K. (1934). *The logic of scientific discovery.* London: Hutchinson, 1959.

Prigogine, I. (1996). *The end of certainty: Time, chaos, and the new laws of nature.* New York: The Free Press.

Real, L. A., & Brown, J. H. (1991). *Foundations of ecology: Classic papers with commentaries*. Chicago: University of Chicago Press.

Reason, P., ed. (1988). *Human inquiry in action: Developments in new paradigm research*. Newbury Park, CA: Sage.

Reis, M. (2000). The ecology of design. *Environmental design and construction* (March/April), 32–35.

Schaffarzick, J., & Hampson, D. (Eds.). (1975). *Strategies for curriculum development*. Berkeley, CA: McCutchan.

Schank, R. (2002). Educating for the future. *Wall Street Journal*, December 27, B1.

Sessions, G., ed. (1995). *Deep ecology for the twenty-first century*. Boston: Shambhala.

Shrader-Frechette, K. S., & McCoy, E. D. (1995). *Method in ecology: Strategies for conservation*. New York: Cambridge University Press.

Sizer, T. R. (1992). *Horace's school: Redesigning the American high school*. Boston: Houghton Mifflin.

Slattery, P. (1995). *Curriculum development in the postmodern era*. New York: Garland Publishing.

Smith, D. W. (1988). *Brain and learning: Perception, consciousness, and empathy*. The Netherlands: Kluwer Academic.

Smith, R. W. (1990). *The role of the transformative teacher: An interpretive inquiry into the possibilities of personal awareness and praxis in authentic educational reform*. Unpublished doctoral dissertation, University of North Carolina, Greensboro.

Stokes, G. (1998). *Popper: Philosophy, politics, and scientific method*. Cambridge: Polity Press.

Thoreau, H. D. (1846). *Walden*. London: St. Martin's Press, 1999.

Ulanowicz, R. E. (1997). *Ecology, the ascendent perspective*. New York: Columbia University Press.

Wallenstein, S. L. (1980). *The reflexive method in curriculum theory: An autobiographical case study.* Unpublished doctoral dissertation, University of Rochester, NY.

Warren, K. J. (Ed.). (1997). *Ecofeminism: Women, culture, nature.* Bloomington: Indiana University Press.

Weber, M. (1946). In Gerth, H. H. and Mills, C. W. (Eds.), *From Max Weber: Essays in sociology.* New York: Oxford University Press.

Weiler, K. (1988). *Women teaching for change: Gender, class and power.* New York: Bergin & Garvey.

Weiler, K. (1995). Remembering and representing life choices: A critical perspective on teachers' oral history narratives. In P. L. McLaren & J. M. Giarelli (Eds.), *Critical theory and educational research* (pp. 127–44). Albany: State University of New York Press.

Williamson, M. S. (1987). *Autobiography as a way of knowing: A student-centered curriculum model using Maya Angelou's "I Know Why the Caged Bird Sings."* Unpublished doctoral dissertation, University of North Carolina, Greensboro.

Willis, P. (1977). *Learning to labor: How working class kids get working class jobs.* New York: Columbia University Press.

Wilson, E. O. (1998). *Consilience: The unity of knowledge.* New York: Alfred A. Knopf.

Zimmerman, M. E. (1994). *Contesting earth's future: Radical ecology and postmodernity.* Berkeley: University of California Press.

A BOOK SERIES OF CURRICULUM STUDIES

This series employs research completed in various disciplines to construct textbooks that will enable public school teachers to reoccupy a vacated public domain—not simply as "consumers" of knowledge, but as active participants in a "complicated conversation" that they themselves will lead. In drawing promiscuously but critically from various academic disciplines and from popular culture, this series will attempt to create a conceptual montage for the teacher who understands that positionality as aspiring to reconstruct a "public" space. *Complicated Conversation* works to resuscitate the progressive project—an educational project in which self-realization and democratization are inevitably intertwined; its task as the new century begins is nothing less than the intellectual formation of a public sphere in education.

The series editor is:

> Dr. William F. Pinar
> Department of Curriculum and Instruction
> 223 Peabody Hall
> Louisiana State University
> Baton Rouge, LA 70803-4728

To order other books in this series, please contact our Customer Service Department:

> (800) 770-LANG (within the U.S.)
> (212) 647-7706 (outside the U.S.)
> (212) 647-7707 FAX

Or browse online by series:

> www.peterlangusa.com